37.95

D0948771

THE HUAWEI WAY

THE HUAWEI WAY

Lessons from an International Tech Giant on Driving Growth by Focusing on Never-Ending Innovation

YANG SHAOLONG
TRANSLATED BY MARTHA AVERY

New York   Chicago   San Francisco   Lisbon   Athens   London
Madrid   Mexico City   Milan   New Delhi   San Juan   Seoul
Singapore   Sydney   Toronto

1 2 3 4 5 6 7 8 9   DOC   21 20 19 18 17 16

ISBN   978-1-259-64305-7
MHID   1-259-64305-0

e-ISBN   978-1-259-64306-4
e-MHID   1-259-64306-9

McGraw-Hill Education books are available at special quantity discounts for use as premiums and sales promotions, or for use in corporate training programs. To contact a representative, please visit the Contact Us page at www.mhprofessional.com.

Library of Congress Cataloging-in-Publication Data

Names: Yang, Shaolong, author.
Title: The Huawei way : lessons from an international tech giant on driving
   growth by focusing on never-ending innovation / Yang Shaolong.
Description: New York : McGraw-Hill, [2017]
Identifiers: LCCN 2016024413 (print) | LCCN 2016031927 (ebook) | ISBN
   9781259643057 (alk. paper) | ISBN 1259643050 | ISBN 9781259643064 () |
   ISBN 1259643069 ()
Subjects: LCSH: Ren, Zhengfei, 1944- | Hua wei ji shu you xian gong si. |
   Electronic industries—China—Management. | High technology
   industries—China—Management. | Technological innovations—China.
Classification: LCC HD9696.A3 C5925 2017 (print) | LCC HD9696.A3 (ebook) |
   DDC 658.4/063—dc23
LC record available at https://lccn.loc.gov/2016024413

CONTENTS

THE *SONG OF HUAWEI* AND HUAWEI'S SPIRIT

The CEO of Huawei, Ren Zhengfei (hereafter Ren), wrote a passionate poem about the company in 1995 that came to symbolize the company's spirit for the next two decades of high-speed growth. It talked about the dream of reinvigorating the Chinese nation, but it also talked about emulating the United States, Japan, and Germany. It dreamed of glory for generations of Chinese, but it also mentioned putting each foot down on solid ground. Meanwhile, China was to learn from American high-tech, Japanese management, and German attention to detail. The *Song of Huawei* encapsulated Ren Zhengfei's ambitions, ideals, and unbending determination but from a unique perspective.

Since the Opium Wars, generations of Chinese patriots have raised a similar call: serve your country by developing it. Rescue it from the ridicule of being the poor man of Asia. Awaken it. Nevertheless, for more than a century China has taken a circuitous path toward such development. Constant wars, internal strife, and external challenges have led it through detours and setbacks. Only in 1978 did a group of companies finally begin to "stride out" successfully into international markets, most notably Lenovo and Haier. In the telecom arena, Huawei became the symbol of China's high-tech.

Ren is generally acknowledged to be the most low-key of any entrepreneur who operates on a global scale. Any world-class businessman must be powerfully motivated to be successful, in addition to having his own way of handling affairs. What motivates Ren? Where does the sense of history that is expressed by his *Song of Huawei* come from, not to mention the sense of destiny? What has he relied on to break through and transcend the extreme limitations that this period of history in China imposed upon him, to the extent that he has indeed led Huawei onto the world stage of high-tech? What can other companies in the world and other corporate managers learn from this company and this man? These are the key issues that this book explores and seeks to explain.

From the time it was a small start-up, Huawei has regarded American companies as the model to emulate. By this, it has meant specifically such companies as IBM, Microsoft, Apple, and Cisco, global leaders in the core technologies of the information industry. These have been the companies that propelled the development of the Internet—in that sense alone, they have made outstanding contributions to humankind. In addition, however, Huawei has regarded the innovative environment in the United States as critical, particularly the mechanisms by which innovations and an innovative spirit can be realized. In the United States, as long as a company or an individual has very good ideas, venture capitalists will consider trying to turn those ideas into reality. The situation in China in the 1980s was radically different. When Huawei was founded, an innovative environment and any mechanisms by which good ideas could be realized barely existed.

When Ren founded the company in 1987, he had both little capital and little "background," or political backing. From almost every perspective, he lacked the innate qualities required to become a leader of a multinational company. In the short space of 20+ years, Huawei nevertheless led its forces into international telecom markets and beat out many superb opponents, including Lucent, Nortel, Alcatel, Nokia, Siemens, Ericsson, and Motorola. In doing this, Huawei established an admirable model for Chinese as well as world companies to follow. Its success has depended in part on the way Ren has led Huawei employees in having an indomitable will to succeed. As chronicled below, the company cultivated an unbending determination and will to fight, but also a spirit of constant innovation. Another key part of Huawei's success has involved its sustained study of America's best managers and entrepreneurs. Having internalized all these things, to a degree Huawei has established itself as a role model for companies around the world.

In the age of global business, if enterprises want to survive, they must constantly innovate with respect to both management and technologies. Otherwise, they will simply lose out in international competition. The successful experience that Huawei has built up over the past 27 years has been won at a cost, which has included not just money but sometimes life itself. This experience now represents a valuable distillation of lessons that can be an asset not just for Chinese companies, but also for the rest of the world. As one who participated in and witnessed this process, I have watched Huawei develop from the inside. As a long-time employee, I also have deep-seated feelings for both Huawei and Ren. In the interests of making these lessons more apparent and broadly known, I decided to summarize Huawei's experience in this book.

After working professionally for nearly two decades at Dell, Huawei, and other companies, I began to think about, research in depth, and then write the story of Huawei. I have used Ren as the primary thread in this story. I approach it in chronological order, while trying to re-create a panoramic view of how this company came to be. The 27-year history of the company to date is marked by milestones and major events, but it is also infused with Ren's sense of mission and responsibility, his transformative spirit. Naturally, I discovered that the difficulties of capturing all this far surpassed what I could have imagined. Major events provide pieces of the story although Huawei's history has not been high-profile. In tracing back through time, I have sought to reflect the realities of the situation as completely as possible. I have taken a scroll of history, deconstructed it, and then put it together again. The process of writing this book, over five years, not only has been slow but has affected my health as well. Nevertheless, if it can provide an example for companies and entrepreneurs around the world, I will feel it has been worthwhile. I believe that the *Song of Huawei* and the spirit of Huawei have extraordinary significance for competition in the global economy. This has motivated me throughout the writing of this book. Just to make things clear, this book only represents my personal point of view and has nothing to do with Huawei.

Now, let us open the windows to our perceptions and see just what enabled Huawei to go from being six people in a small workshop to becoming a multinational giant with 190,000 employees and annual revenues that are in the neighborhood of USD 62 billion.

Yang Shaolong
October 2015
Qingdao, China
yangshaolong1965@163.com

INTRODUCTION

K Street runs through the northern part of Washington, D.C. At its eastern end, it connects with Capitol Hill, and at its northern end it runs into Georgetown. Although it seems no different from any other street in America from the outside, the very name K Street is in fact a kind of huge ad. It declares, "People in the White House can only tell you what just happened. We, on the other hand, can help you change what is going to happen."

America operates with a separation of three powers, namely, the Supreme Court, Congress, and the White House. Those may indeed comprise the three primary centers of power, but there is a reason K Street has been called the "fourth center of power." More than 100 think tanks are concentrated here, with their fingers on the pulse and with very complex "backgrounds." They use all kinds of channels, lobbying groups, PR companies, private organizations, international headquarters, and rumor mills among government officials and members of the judiciary to influence and change domestic and foreign policy in America. Their customers come from every part of the world and every walk of life. You can find government people, arms dealers, oil magnates, and anyone else who hopes to change American policy.

The day after Valentine's Day in 2010, a particular PR company on K Street received several guests from Asia. After discussion, this company accepted their commission. Their role was "to make government officials maintain an objective and fair attitude and business stance when Sprint Nextel, the third-largest telecom equipment operator in America, a company that buys several tens of billions of dollars' worth of goods and services every year, calls for tenders for telecom equipment."

The guests were from a company in China called Huawei.

Who Is Huawei?

With this K Street public relations company serving as go-between, as the needle and thread that tie the interests of parties together, Huawei was introduced to a suitable partner. This was an American telecom company called Amerilink. The PR company had chosen this partner carefully. Amerilink soon invited a number of former senior government officials to participate in a lobbying team. To ensure that Huawei would get the tender, Amerilink announced at the appropriate time that it would be jointly bidding with Huawei. At the same time, it recruited seven high-ranking technical R&D people who were familiar with Sprint Nextel's product needs and functions and who were therefore quite well paid for their advice.

Even with this massive array of force lobbying on its behalf, however, Huawei found that things did not work out well.

In December 2010, Huawei placed first in terms of its technology, business terms, and equipment functionality in the tenders for 3G equipment worth USD 5 billion. Since this business had "national security" implications, however, Huawei did not get the order.

The company had encountered this kind of Waterloo in America several times before. In December 2007, Huawei unsuccessfully joined hands with the private equity fund Bain Capital to try to purchase the Internet equipment firm 3Com. On July 16, 2010, Huawei joined in trying to purchase the wireless network assets under the banner of Motorola. In July 2010, Huawei tried to acquire 3Leaf. All these attempts met with failure due to "national security" issues.

The reason is simple. Although 71 years old in 2010, the company's CEO, Ren, had once been an officer in the People's Liberation Army of China. What did the company he founded have to do with the Chinese government and with the Chinese military? Who, after all, was really running Huawei? These questions were key issues leading to Huawei's failures.

Investigating Ren's "Background"

For many years, a number of countries and governments have launched all sorts of investigations into Ren and Huawei, as Huawei has begun to emerge on the international scene. They have done this in the name of national security.

In fact, it is quite common for people who served in the military to found or to run companies. In March 2003, when Cisco and Huawei engaged in a significant lawsuit about intellectual property rights, 3Com announced it was setting up a joint venture with Huawei and as a result came under media scrutiny itself. The board of directors of 3Com made quite an objective comment on the issue: "If military service precludes someone from being a CEO, over half of America's CEOs would not have a job today."

Nevertheless, there are indeed reasons for the doubts and concerns held by some countries and governments. As the founder of a company, Ren was out of the public eye for a long time. He would not be interviewed by anyone from Chinese or international media. This desire to stay out of the limelight is not only unique in China, but highly uncommon around the world. It was hard for people not to suspect that there was something to hide. Surely there was some secret that could not be spoken of openly. Moreover, before Huawei moved in the direction of international markets, Ren had said such things as this: "The main entities engaged in international competition in the future will be corporations, not governments." "The competition between one country and another is in fact a competition between major corporations." Since he took such a clear stand on this, it was easy to endow his words with special meaning. It was easy to think he might be speaking for China as well.

Suspicions have bred more suspicions. Huawei started out as a small operation with RMB 20,000 in registered capital. Even in the United States, it would take nothing less than a miracle for such a company to beat out multinationals within a mere 27 years and earn annual revenues in excess of USD 48 billion. Meanwhile, the United States has a wealth of innovative talent, while China has a very thin substrate on which to grow a business. Ren was understandably questioned and blocked by both countries and governments. Despite this, however, international media have generally valued his accomplishments highly. The publicity has shown respect for the man himself as well as the company he founded.

In April 2005, *Time* magazine chose Ren to be among the 100 most influential men on the globe, in company with such famous IT representatives as Bill Gates of Microsoft, Steve Jobs of Apple, and the founders of Google, Larry Page and Sergey Brin.

In December 2008, *Business Week* chose Huawei to be among the 10 most influential companies on earth. It thereby joined the ranks of others chosen at the same time, such as Apple, Google, and Toyota.

In February 2009, the World Intellectual Property Organization announced that Huawei had filed for 1,737 international patents in 2008, beating out Matsushita, Phillips, Toyota, and others to become the world's "patent champion."

In July 2010, *Fortune* magazine announced that Huawei had successfully joined the ranks of the Fortune 500, with USD 21.8 billion in revenues. It was ranked as number 397, but was the sole Fortune 500 company that was not listed on a stock exchange.

When this news broke, it generated a global media commotion. *Businessweek* followed it up with an editorial that said, "Huawei, based in Shenzhen, has already become one of the world's largest telecom equipment manufacturers. Based on its patents and innovations, not only has it become the model for China's new-style enterprises, but it has become a leader in global change." The consulting company Ovum, which focuses on researching telecom developments, said that Huawei had already gone from being a "Chinese vendor" to becoming a "global giant."

Given all this, what kind of person is Ren, in fact? What does the company he founded have to do with the Chinese government and the Chinese military?

To unravel this mystery, you have to lift the veil of secrecy that covers Ren. You have to analyze his thinking, his character, and the very source of his DNA. Only once you have read Ren accurately can you understand the drama that is Huawei.

PART ONE

FOUNDING THE COMPANY

Only if you dare to dream of doing things can you begin to do them; only if you dream of creating revolutionary change can you ever get good at it; only in the midst of such revolutionary change can new opportunities arise.

—REN ZHENGFEI

This statement was an article of faith for Ren Zhengfei, but it also described the road to success for Huawei.

CHAPTER ONE

REN ZHENGFEI'S UPBRINGING AND THE ORIGINS OF HUAWEI

Mental trauma and a tough life in terms of material needs provided the opportunity to become more mature in later life.

—REN ZHENGFEI

Although he was not born there, Ren Zhengfei's ancestral home is in Zhejiang province. He was born in an impoverished mountainous part of Guizhou province, in a region called Anshun.

His father, Ren Mosun, was born into a wealthy commercial family that dealt in the ham business in Jiangjinhua, Zhejiang province. Ren Mosun's native intelligence, hard work, and appreciation for studying enabled him to pass the exam for college in 1929 with high marks. He then entered the economic department of what at the time was known as the Minguo University of Beiping. By 1946, he was living in a mountainous part of Guizhou province that was designated an ethnic-minority region. There he built two schools, one a middle school called the Zhenning ethnic-minority middle school and the other a specialized training school called the Dujun ethnic-minority normal school. For the next 40 years, Ren Mosun served as the principal of first one and then the other of these two schools.

The mother of Ren Zhengfei (hereafter referred to as Ren), Cheng Yuanzhao, was a native of Guizhou province. Although she had only a high school education, she became a professor of mathematics under the influence of her husband and through her own self-study.

Accompanying her husband in the remote and impoverished mountainous area of Guizhou, she dedicated a lifetime to helping poor children. She was the classic model of a traditional Chinese woman, a hard worker with tremendous stamina, kind-hearted, industrious, and thrifty in managing her household.

This family environment of teachers and education had a profound influence on Ren's sense of values.

Youth

With six younger siblings, Ren ranked eldest of the seven children in the family. All nine members of the family depended on the meager salary of teachers to make ends meet.

As the children grew up, not only did they need to be fed but also they needed to be educated. Clothes were cut down for each succeeding child, but the family expenses were still hard to manage. Ren remembers his mother worrying about the 2 or 3 RMB needed for tuition every semester—at the end of the month, she would have to ask neighbors for money to buy food. What's more, she often would have to go to several houses before she was successful in getting that money.

The children slept two or three under one quilted blanket. Beneath a ragged sheet, rice straw was used for bedding. The family had no stove—a fire was built in a hole dug into the floor, and it was used for cooking as well as for warmth.

Given the impoverished circumstances, Ren never wore an undershirt until after college. On the hottest days, he simply wore the same thick overshirt. When his classmates told him he should ask his mother for an undershirt, he responded that he didn't want to do so since he knew she couldn't manage to buy one.

After he had passed the entrance exam to Chongqing Engineering College, on a one-time basis, his mother gave him two undershirts. He has said that at the time he just wanted to cry. With this gift, his little brothers and sisters would be even more deprived. Since he had no sheets, his mother found one that had been discarded by another student at the school, then carefully mended and washed it. This one sheet was to serve Ren for the next five years as he went through college.

Between 1959 and 1962, the food supply became extremely limited throughout the country during a calamitous three years of intensely dry weather. Life became even tougher for Ren's family, with its many children. In his third year of high school, Ren's chief ambition was "to have an entire roll to eat." During years that were intensely difficult for all Chinese people, the selfless nature of his parents was to have a profound influence on Ren. He remembers this period of hardship well. "I was 14 or 15 years old during the worst time. I was the eldest. All of my siblings were younger and could not understand what was going on. My parents could easily have eaten another mouthful of rice without anyone knowing about it, secretly, but they did not. If they had, one or two of my siblings would not be alive today."[1]

"When it was getting close to time to take the entrance exam for college, I was so hungry I could scarcely study. I would mix together rice husks with some leaves and roast it. My father caught me at this a few times, and it hurt him to see it. Rice was stored in an earthenware jug. We never took more than what was allowed. If we had, again, one or two of my little brothers and sisters would not have lived."

This very brutal childhood gave Ren a strong sense of the difficulties of this world at an early age. He had an acute understanding of what it meant to "keep on living." The teaching of his parents, as well as their example, inculcated in him a frugal and selfless attitude. This virtue was to run through his entire life.

In 1997, after 10 hard years as a start-up, Huawei went from being a small workshop operation to gradually becoming dominant in telecom manufacturing within China. To develop the Latin American market, Ren now decided to invest USD 30 million in Brazil in setting up a joint venture. After the signing ceremony for the joint venture was over, the head of the Brazilian side of the venture invited Ren to travel through

---

[1]During the three years of natural disaster, as officially designated, several tens of millions of people died of starvation or starvation-related causes. This shaped the psychology of many people who are now China's older generation, including Ren.

the tropical rainforest part of the Amazon. Before leaving home, senior people at Huawei bought fancy shoes for the trip without thinking twice about it, flashy Adidas and Nike. Ren instead purchased a pair of cheap rubber shoes. Once everyone got back from the Amazon, the others threw away shoes that were now moldy and covered with mud. Ren simply rinsed his off, hung them out to dry in the sun, and then took them back to China with him.

Not long after, in 1997, China's universities began requiring tuition fees of students. Policies granting tuition subsidies did not keep pace with this, so Ren proposed that Huawei establish a fund for so-called *hanmen* students, those from poor and humble families, and the company contributed RMB 250 million to this fund for impoverished students. The incident provided Huawei's senior management with a good lesson—Ren was tight about his own needs yet open-handed when it came to public welfare. From then on, senior management tried not to be openly extravagant. They chose the cheapest alternative if the thing was being bought for their own personal use, and they chose things that would last the longest.

In addition to frugality, another notable feature about Ren has been that he does not emphasize his own personal wealth. He has been willing to share the results of Huawei's "struggle" with employees. This has allowed him to attract and hold outstanding human talent and to make Huawei "great" in the process.

In August 1995, a former vice president of the Stone Group joined Huawei, a man named Li Yuzhuo. Not long after, Li Yuzhuo introduced Ren to the chairman of the board of Sichuan's Stone Group, Duan Yongji. The focus of their discussion during a meeting in Shenzhen was the reform of ownership systems within China, in order to create shareholding systems. This subject was quite sensitive among China's state-owned enterprises at the time. Once Duan Yongji realized that Ren held only slightly more than 1 percent of all shares in Huawei, and that senior management held less than 3 percent, he asked Ren, "If you all only hold 3 percent of shares, have you realized that you could be overthrown someday by a group of people that gets together to oust you?" Ren's answer astonished everyone in the room. "If they can oust me, I think that would be just fine. That would be a clear indication that the company is mature."

Li Yuzhuo expressed his conclusions about this exchange later. "In that moment, I understood that Ren had taken over the banner of '*minying* companies'[2] from Duan Yongji. This was a banner that the Stone Group had been able to wave for the past 10 years."

Looking back over the course of Huawei's growth over 27 years reveals the reason Ren has been able to consolidate and hold together a group of superlative people who fight for a common goal: it has been his willingness to disperse wealth to others, his courage in delegating authority, and his selfless nature. This is acknowledged by many to be the key reason Huawei has gone from being six people to what it is today, a multinational with 190,000 employees. The great majority of those employees feel a strong personal attachment and loyalty to the man.[3]

College

In 1963, Ren passed the examination with flying colors to get into the Chongqing Institute of Civil Engineering and Architecture. Like the majority of students, he was extremely diligent. Among various courses, he excelled in English in particular. Although he majored in engineering, he showed talent in the humanities and in logic.

When he was in his third year of college, many institutes of higher education stopped holding classes due to the eruption of China's Great Cultural Revolution. College students began to throw themselves into the conflicts between different internal factions. During this period of turmoil, Ren threw himself into studying. He read the entire *Collection of Higher Mathematics Studies* from start to finish, twice. He borrowed mimeographed books from professors, and he began to teach himself "electronic calculating," numerical-control technology, automated controls and logic, philosophy, and three different foreign languages. In addition, he systematically read through and analyzed the four-volume *Collected Works of Mao Zedong* as well as *On War* by Clausewitz.

---

[2]The ownerships of private companies (no state-owned companies) can belong to the state or individual persons. Huawei is the second kind, it is privately owned.

[3]Ren's efforts to get beyond this adulation and this mythologizing are recorded below.

During this "period of turmoil," Ren adhered to his program of self-study and amassed considerable knowledge. Once he joined the army, this enabled him to shine among numerous other talented people and to display technical expertise in particular. A number of his technical inventions received "#1 awards" in the army, and two of these filled specific national needs.

In 1987, after Ren founded Huawei, he set up a series of incentive programs and ways to evaluate performance that were successful in pulling together a critical mass of technical talent. These were based on his respect for technical ability and on his sincere respect for knowledge. The incentive programs allowed Huawei to set out upon the "track" of high-tech. The programs were encapsulated in various catch phrases, such as these: "Labor, knowledge, entrepreneurship, and capital are what create the value in a company." "Employees are the greatest wealth of Huawei. Respect knowledge, respect individuality, and don't simply go along with key individuals, but fight together as a group."

Since Ren had thoroughly read and absorbed such works as the *Collected Works of Mao Zedong* and *On War,* by Clausewitz,[4] he later applied the lessons learned to the management of Huawei. In many ways, these enabled the company to survive in an extremely hostile environment. They enabled Ren to "use the small to conquer the large," to "use the weak to beat out the strong," and to use historic examples to grow the company to be big and strong.

"It may well be catastrophic if one's life is too easy. One can notice that life's setbacks often count as blessings, rather than as curses." "If I had not been able to see through to light at that time, and if I had wasted that time, I would not be where I am today." Ren has often reiterated such thoughts in conversations with employees.

Even though Ren came up against a Cultural Revolution that occurred during his college period, in the end he was indeed able to attend college. Despite having received this education, he has never relinquished his pursuit of knowledge—indeed, he could be called a disciple of knowledge.

---

[4]In Chinese, the title of the book is *The Art of War*, which echoes the Chinese work by Sunzi, also called *The Art of War* [*Sunzi bingfa*]. *On War* was written by Carl von Clausewitz in 1832.

In the end, he has used knowledge to change his own destiny as well as that of Huawei. Through systematic and intensive study of such works as *On War*, he has both broadened his perspective and matured in his thinking about politics and ideology. After college, Ren was no longer purely a bookworm, but these things helped lay the theoretical and ideological foundations for his becoming a world-class entrepreneur.

Serving in the Military and Firm Convictions

In 1974, after China's position in the United Nations was restored, China entered a period of economic development. The country also began to import large quantities of equipment from the United States, England, and France, in order to resume production. To this end, it increased the number of troops engaged in "capital construction projects," that is, in the building of basic infrastructure. Ren now joined the ranks of these "troops" in the capacity of a "technical soldier." As a technician, he was put in the military's Engineering Corps.

Ten years of life in the army did not in fact bring Ren a great deal in the way of glory or professional accomplishments. However, the army honed and disciplined his powerful ambition, sense of mission, and sense of responsibility. Ren's prestige and influence in Huawei relate less to the fact that he is "the founder" than to his ability to serve as an example and thereby get people to follow him. He holds power by being selfless, fearless, and willing to lead. This creates a force that is unlimited and that is therefore also invincible. It also represents the single largest way in which military life influenced Ren.

Ren does not own a car. He pays for his Huawei phone out of his own pocket. He takes responsibility for and punishes himself for mistakes made in documents that he has signed. When any employee in the company uses his position for private gain, Ren rules on the matter immediately and decisively, no matter whether that person is an "original founder" or part of the company's "new aristocracy." Examples of this are legion. In July 1997, Huawei was forced to have a reduction in force, which was conducted according to its system of "washing out those at the end of the line." To avoid later grievances among people who had been fired, Ren met with each person individually and encouraged him to file a suit or grievance if he had different views on things. One time, he found out that an employee in a production department felt he was being fired because his own superior had not followed adequate quality control procedures. This led to the production of spare parts that then had to be thrown away. When it came time to "squeeze out" those who were to be fired, Ren called for the manager who had been responsible for this lapse in order to confront the issue in person. Ren became absolutely furious when he learned that the employee had been correct. For the first and only time, he gave the manager a blow on the ear. He then told the man that he was fired.

As a result of this experience, Ren personally drafted a system called "forums for employees who are put in an adverse position." The contents of the forums are made public in a newspaper called *Huawei People* and in the journal called *Better Management*. The system is used to warn senior cadres: there are no privileged people in Huawei, no "special rights." "We forbid the exercise of power in order to seek personal gain, and we forbid the use of force to intimidate the weak."

Although Ren has dealt severely with senior cadres inside the company, he can be extremely considerate of common employees, particularly when it comes to their personal lives. Under that ferocious exterior, he has a sensitive and gentle nature. He often uses the old saying about how generals should treat their soldiers as they would their own sons, care for them as their own brothers.

In early 1997, a secretary in the marketing department of Huawei, Yang Lin, was killed in a car accident in Hainan. After receiving this shocking news, Ren wrote a eulogy for her in his grief. He described the selfless contributions she had made to the company in its early period, and he called on all employees to learn from her example.

A sales and marketing employee in the marketing department had long known he had liver disease. Nevertheless, he persevered in the northeastern sales territory, in order to pull in orders for the company. In the end, his condition deteriorated and he died. After this, Ren wrote a profoundly moving letter to employees called, "Do Not Forget Our Heroes."

After 2000, as Huawei's overseas business swiftly expanded, Ren spent most of his time making the rounds of overseas Huawei organizations, in on-site inspections. In Africa, he discovered that the prevalence of malaria-transmitting mosquitos was a huge problem. When visiting the American Consulate, he learned about a device used specifically by the American military for repelling mosquitos, and he purchased this. After returning to Shenzhen, he experimented with it himself on hot humid nights. Once he discovered that it was indeed effective, he promoted its use in more than 100 Huawei locations overseas.

Ren understood quite clearly that a powerful, ironclad army is built up through love, not through floggings.

The success of an entrepreneur must necessarily be driven by a powerful spirit. At the same time, the sources underlying this driving force must come from unwavering beliefs or convictions. Huawei University was established in 1998, and from the start its emphasis was on a military approach to business. "The market is a battlefield" was the theme. The first book that Ren recommended to the human resource department of Huawei was called *West Point Academy as Our Guiding Spirit*. He told the human resources department that it should focus on a philosophy of management that treated human resource training as one would treat the indoctrination of troops. Meanwhile, Ren modified the three "great beliefs" in which West Point trains its soldiers. Instead of adhering to "responsibility, glory, and nation," he added "our endeavor," meaning that Huawei employees were to focus on "responsibility, glory, *our endeavor*, and the nation." What's more, Ren had a huge banner with a slogan on it hung in the most visible place at Huawei University. This then became the "oath" that Huawei employees were to keep in mind at all times.

Wu Jianguo served as deputy head of the human resources department at Huawei during this period. He recalls that Ren repeatedly emphasized to him that the mission of military personnel was to "safeguard the home and protect the nation." In contrast, an entrepreneur had to insert this new thing, "endeavor," and the endeavor—the long-range ambition—was to become a world-class enterprise. Military men could sacrifice their lives on behalf of the national interest. Entrepreneurs should similarly sacrifice everything in pursuit of this "endeavor," including their own personal wealth, interests, and an easy life. What one can accomplish depends on one's breadth of vision, one's mental horizons. Not only does this require reaching for ideals, but it particularly involves willingness to sacrifice oneself as well.

In this, one can see the sense of values that Ren has consistently maintained, and one can also recognize his convictions.

Founding Huawei in a Time of Hardship

In 1984, as China shifted into wholesale "economic construction," the country also began cutting back on its military forces by the millions. The Capital Construction Engineering Corps was reorganized and dismantled. Ren switched occupations. Formerly in the army, he now became the manager of an electronics company in Shenzhen that was under the jurisdiction of the Shenzhen South Sea Oil Corporation Group.

Ren had no commercial experience from his college days or throughout his army experience. He also had no expertise in managing a company, so in the course of handling business projects he was soon cheated out of the massive sum of more than RMB 2 million by a trading company. At the time, this was an enormous amount of money in China. Because of this fiasco he was forced to resign from the group. There was no job waiting for him, and his path in life was now being forced in ever-narrower directions. To survive, he and six friends pulled together RMB 20,000 and founded what they called the "Shenzhen Huawei Technologies, Co., Ltd."

When his parents heard of this series of changes in their son's life, they immediately traveled from Guizhou to Shenzhen to help him. The whole family squeezed into his small home, some dozen square meters in size. They cooked on the balcony. Huawei was founded under extremely difficult conditions, since the loss of RMB 2 million now became a debt that Ren was obliged to pay, and this pressed on his conscience. His parents had good reason to worry about him. His father brought low-quality tobacco all the way from Guizhou and he smoked incessantly. Every evening, his mother would go out shopping as the markets were closing, to buy scraps of fish, leftover shrimp, old vegetable leaves, and so on. She held back all the retirement funds she was getting in order to "rescue" her son.

This period of time was undeniably difficult for Ren. Nevertheless, the many low spots in people's lives force them to be creative and to confirm that "China will find a way." Even at this earliest time, one can recognize Ren's tremendous ambition.

Although it wore the title of a tech company from the outset, what Huawei really did was trade. Since none of the six had ever had any commercial experience, they took a while to figure out the direction in which they wanted to go. Essentially, they did anything that made money. They sold balloons, fire alarms, and at one time they even sold dietary supplements.

At one chance meeting, Ren learned from a friend that there was an enormous demand for small telephone exchange switches used in such industries as mining companies, hotels, public security, and so on. After looking into it, he decided to be an agent for Hong Kong's Hongnian Company, for the HAX switches. China already had more than 200 companies dealing in small switches at the time, but this included all kinds of small-fries among the bigger fish. Some of them did assembly and production themselves, with low-quality results. Others imported directly from overseas, but in that case the price was exorbitant. Relatively speaking, the HAX switches that Huawei was now representing were better in quality than domestic equipment, but also lower in price than imported equipment. Moreover, Huawei emphasized post-sales service. It opened up this market so quickly and orders came in so massively that Hong Kong's Hongnian Company had trouble getting enough products to meet the demand.

With a market as explosive as this, Ren quickly asked Zheng Baoyong to organize an "attack" team. This team then soon came out with what was called an HJD48 switch. Although it only had 48 "access points," it was quite profitable since the technology had been the result of independent R&D. Within the short space of one year, sales exceeded RMB 100 million, and Huawei grew to have more than 100 employees.

The matter might have ended there. The situation might have resulted in the usual story of Shenzhen: distribute the dividends, play the market shares by manipulating them upward, then get out of the country. Instead, Ren, who now had an extremely acute understanding of business, discovered an even greater opportunity. This opportunity was in what are called program-controlled telephone switches.

China's "Don Quixote"

In the early 1990s in China, a massive advertisement was hung in the most visible parts of China's streets and alleys. It advertised "programmed" or "automatically controlled" telephoning. It said that one could actually dial long-distance and be connected automatically. In this ad, a map of China was presented in seven different colors, which showed how China's market for program control switches was monopolized by companies from other countries. These were Japan's NEC and Fujitsu, America's Lucent, Germany's Siemens, Belgian Bell, Sweden's Ericsson, and France's Alcatel. They had "carved up the pumpkin" into their own spheres of influence. This later became known as being carved up by "seven countries and eight systems." The result of this monopoly situation was that you had to pay between RMB 3,000 and RMB 5,000 to install a telephone in China, plus you had to wait in line for at least half a year. Meanwhile, in a population of some 1.2 billion people, only 0.5 percent had landline connections.

In 1992, the 38-year-old head of the Army's Information Engineering Institute, Professor Wu Jiangxing, successfully developed a G-4 numerically controlled switch, Digital SPC Exchange, destroying the myth that the Chinese would never be able to come up with its own large-scale switching equipment. This news shook Ren to the core. Several of the founders had blocked his attempt when he had told them earlier that he wanted to start researching how to produce switches.

The founders' reasons were very practical and also quite obvious. Here is an example: In 1988, when the chairman of the board of Vanke, Wang Shi, openly began selling shares on the streets of Shenzhen for RMB 1 per share, nobody paid any attention. Wang Shi sent his secretary over to talk to Ren. After explaining how a shareholding system worked in detail, and what national policies governed the system, Ren mobilized his relatives and friends to come up with the money for a one-time purchase of RMB 200,000 worth of shares. After Vanke went public, these shares soared in value. By 1992, both the share price and the value of the gifted shares had quadrupled.[5] Ren and his partners naturally made a bundle. This first testing of the water had been very positive. If people just focused all their energy on the Chinese stock market, was the thinking of the founders, then the sky was the limit. Why work too hard? Moreover, at that time Shenzhen was in the midst of a period of freewheeling *dao-mai*, which involved buying things at controlled state prices and then reselling them for market prices. If you had any intelligence at all, you could make large sums of money. In contrast, going into making switches was a very uncertain business. A private entrepreneur engaged in a tech field, making telecom products, was not what the cofounders wanted to hear about. If they put hard-earned money, several million RMB, into R&D, who could guarantee they would even get back their sweat capital?

Several founders therefore begged Ren not to take the risk. "Whatever you do, do not take any risk." Ren indicated that he understood this very well. In those opportunistic times, the truth was that it was safest to leave your wallet at home. Ren later recalled this time with some emotion during a meeting to "drill and train" the marketing department at the end of 1995.

"Huawei was founded in Shenzhen during its heyday of *dao-mai*. People who did anything tech were considered mad. We were fools. We were out of sync with our environment, and therefore subject to intense hardship as a start-up. You cannot imagine it."

"Comrade Zheng Baoyong started with the 40-switch. He served both as a worker in actual production and as a frontline person, installing phone lines. An outsider, whose business was lasers, he courageously groped his way forward and came up with the idea of targeting the most advanced level of technology in the world, and he even dared to surpass ATT's 5G machines. Then he ran into this fool Ren! These two guys were like a modern kind of Don Quixote.

Reverberations from a Trip to America

Once he had decided on the goal of "overtaking" other countries in this particular technology, Ren realized that he had better examine the United

---

[5]For the first time to enter the stock market, Ren earned 20 million yuan, in 1992, which is a huge sum of money, he shows amazing talent.

States from a closer distance. He needed to see just how much of a gap he had to overcome.

In the fall of 1992, he led a team of his backbone staff to the United States. For the very first time, he and the others set foot on American soil. Over the course of just one week, the team traveled several thousand kilometers in a rented long-distance van, visiting Boston, Los Vegas, Dallas, New York, and Santa Clara. This fleeting glimpse of America could play no more than the slightest role in educating him about its modern corporate ways, but in terms of perceptions, it provided a massive jolt to his system. It directly provoked and confirmed his determination to pursue high-tech.

In Dallas, he visited the headquarters of Texas Instruments. When Ren learned that the grounds covered hundreds of acres, he was astonished and declared he would like to see the factories. The head of the Asia Region told him, however, that there were no factories there. All the factories were spread out over more than 100 countries around the world.

At the IBM headquarters, Ren had originally planned to see this most representative of American giants in just half a day. Once he came through the main gate of IBM, however, he found out that it occupied so much ground that you could not see it in half a day even if you rented a taxi to drive around it.

In New York, they visited Central Park and were astonished to find that it contained old woods. After walking for what felt like a long time, they looked on a map and realized they had covered just a small corner. This visit to New York struck a chord in Ren. It made him think about and be concerned about China's own situation and its future development. "Our land mass is roughly the same as that of the United States. But Tibet and Xinjiang take up half of that. The Yunnan high plateau takes up another portion, so what you have left is less than half of the United States, on which we have 1.2 billion people. With no money for education, and low cultural understanding, we blindly breed more people even as every inch of land is used for something, any corner of remaining soil is dug out. How could we possibly leave any space for woods or grass?"

Before leaving New York, Ren took the remaining hour to pay a visit to the Museum of Natural History. As he strode through the museum, he discovered that there were lifelike marvels of nature from every part of the world, and that they had been displayed and organized according to natural laws and in the most harmonious way. He felt as though he had returned to nature and to primitive society. After an hour, they had not even covered one corner of one story of the five-story building. As classes of American schoolchildren toured the museum, lost in wonder, he exclaimed to his companions, "The more prosperous you are, the more you can advance science and technology. The more advanced your science and technology, the more you can focus on education and produce generations of talent, and the more talent you produce, the more your economy prospers. You move into a virtuous cycle. In China, most of our rural students are studying in precarious buildings. People don't even think about engineering when they build them. I have to say that concern for the nation and for our people is like a shadow over my heart."

He also noted, "Who knows what inventions are still to come? There are tomorrow's atomic bombs, tomorrow's space flights. Throughout this process, the United States is going to endure. This country is going to be around for a long time."

The last stop on this trip was Silicon Valley in Santa Clara

Santa Clara County is where Silicon Valley is located. It includes San Jose, Palo Alto, and Mountain View, and when Ren visited it, it had a population of around 1 million people. (It is now 1.8 million.) A great deal of America's high-tech is located here as well as the R&D centers of the most advanced, sophisticated parts of the industry. If you were to look at California's economy apart from that of the United States, California's GDP alone would rank eighth in the world.

That evening, in a hotel in Silicon Valley, Ren tossed and turned, unable to sleep. All he had seen and heard on this tour flickered like a movie through his mind. At dawn, he stood on the balcony of the hotel room, looking out, and saw the lamps of Silicon Valley still brightly lit. Countless numbers of people in Silicon Valley were spending sleepless nights in a race against time. This, he felt, was what had produced space flights, silicon chips, supercomputers. This had created the great wave of the electronics industry that swept the world, and this thereby had created the prosperity of the United States. He later wrote that this moment sparked the awakening of a powerful dream in his mind. In highly emotional terms, Ren wrote the following in what became known as his "Notes on a Trip to America":

Silicon Valley made the deepest impression on us. Every vein in my body pulsed with the shock of what I saw. We recognized that our own methods of R&D are extremely backward. Our levels of R&D management are extremely primitive. Our efficiency is far behind that of developed countries. We simply have a long way to go.catch up. One point is worth celebrating, which is that the quality of our employees is not any less than that of American companies. We will therefore be able to catch up with America. One thing is of ultimate importance, however, and that is that we must improve our management.

Once I returned, our company decided to buy some rooms in the heart of Silicon Valley and set up an R&D center there. We decided to take the results of our Chinese-side R&D and half-baked products and improve the design in Silicon Valley. Then afterward we could move them back to Shenzhen for production. We also decided to register a company in Silicon Valley. We applied for a permit for a wholly Huawei-invested Lanbo Technologies Company, Ltd.

On the flight from Los Angeles back to China, Ren enjoyed a movie starring Tom Cruise and Nicole Kidman. Two young people from England with dreams and ambitions endure hardships on their trip to the west and fall in love in the process. The young man almost loses his life in order to get land. As the movie ended, Ren again jotted down his feelings. "The prosperity of the west was created out of a heartless desert. Americans went through a great deal more hardship than what we are experiencing today. America's prosperity was earned through the sacrifices of many generations. We should learn from their indomitable spirit."

This trip to America sent Ren back to China bathed in powerful emotions and also steeped in the determination to learn all he could from the United States.

Dividing the Market into Thirds

On return to Shenzhen, Ren took no time for rest. Instead, he immediately brought together his R&D staff to hear about the trip, and he made sure they had an acute understanding and appreciation of what he had seen and learned. He decided, moreover, to set up an "attack team" with Zheng Baoyong and Li Yinan as its core members. Their task was to target ATT's 5G machines and to concentrate all possible resources to make powerful breakthroughs in program-controlled switching technology.

Dreams can be a powerful motivation, but reality unfortunately can still intervene when it comes to realizing those dreams. This was particularly true with respect to the massive sum of money needed for the R&D required to make dreams come true.

In 1992, to tamp down the overheated economy, China's government ordered banks to tighten up on bank loans. The People's Bank of China announced that it would strictly control bank credit, starting in 1993. At the same time, it would conduct a "rectification" or reorganization of the banking industry. This caused several commercial banks in Shenzhen to shut their doors to Huawei. The final "frost on the snow" came when, after several years of high-speed growth, the market in China for small-scale switches moved toward saturation in 1993. The R&D investment required for program control switches remained a bottomless pit, however, and it quickly put Huawei in trouble with cash flow.

Everyone in the company was in a panic about what would happen. It was in this difficult patch, besieged on all sides, that Ren began instituting a system of employee-owned shares within the company. By using what he called "distributing responsibility as well as profits among employees," he gradually reduced the anxieties and stabilized people's psychology. He raised RMB 79 million in funds from external sources, at an interest rate of 33 percent per annum. This sum was able to resolve the bottleneck Huawei had been facing in its R&D. It enabled Huawei to survive when it was *in extremis*.

On April 19, 1993, to give employees that indomitable spirit, at the annual meeting of the people involved in developing Huawei's program-controlled switches, Ren stood in front of a window and quietly said, "If this effort, this round of R&D, fails, I am going to jump out of this window. You all can find your own way out." Given the dramatic tone of this declaration, one can see both the enormous pressures Ren faced as well as his determination to burn the bridges behind Huawei and force it to fight.

In the following battles, with their back to the water, Huawei's people ran faster, worked harder, and paid a greater price than anyone else—to make breakthroughs in technology, grab markets, and succeed. When they were tired, they lay down in the office to take a short rest, which is the source of the phrase used to this day in the company about a "cot culture."

"The moment one definitely commits oneself, then Providence moves too. All sorts of things occur to help one that would never otherwise have occurred. A whole stream of events issues from the decision, raising in one's favor all manner of unforeseen incidents and meetings and material assistance, which no man could have dreamed would have come his way. Whatever you can do, or dream you can do, begin it. Boldness has genius, power, and magic in it. Begin it now." This statement is attributed partly to Goethe, partly to others, but it certainly described the state of mind of Huawei people at that time.

After countless failures, and after the investment of upward of RMB 100 million, a group of students who had just graduated from college achieved a measure of success. Under the most primitive conditions for doing R&D, they experimented their way forward and ultimately put out program-controlled switches that carried their own intellectual property rights.

On October 25, 1994, Huawei was the one and only Chinese manufacturer of telecom equipment to be exhibiting at the First Beijing International Telecommunications Exhibition. In the exhibition hall, a red flag with five stars on it was hoisted aloft amid a sea of foreign equipment. It was at this exhibition that Ren, heart pounding, declared to Chinese guests, "In 10 years, the market for telecom equipment will be divided in thirds among three companies, Siemens, Alcatel, and Huawei."

For a small shop that virtually started out in a thatched hut,[6] this statement was no less than a declaration of war against the global community of telecom giants. When we look back on these past events from today's perspective, we can only imagine the emotions that animated Ren back then. Who could have expected that this Chinese tech company, *ge-ti-hu,*[7] would in the next 20+ years overturn the market structure of world telecom? Even more, who could have known that the company would become an unmitigated nightmare to telecom giants in both Europe and the United States?!

---

[6]This is the equivalent in China for "started in a garage."
[7]The term *ge-ti-hu* here, literally a sole proprietorship, emphasizes the nongovernmental nature of the company at that time.

CHAPTER TWO

THE MARKET SHEDS NO TEARS

The market does not have time to wait for us to grow up. It isn't our mother. It has no patience and it has no mercy.

—REN ZHENGFEI

Despite his declaration about dividing the market into thirds and taking one-third, Ren Zhengfei still had to contend with the longstanding monopoly that foreign companies enjoyed over the China market for program-controlled switches. He was full of ambition, but after R&D came up with the products, he still needed to break this monopoly.

In 1993, Huawei's sales and marketing department contained just sixteen people. This team crisscrossed the country, scouring the market for small-scale switches. It went after such customers as hotels, the public security department, and industrial and mining companies. Huawei was totally unknown to the telecom market itself, however, which was a very powerful sector controlled by national policy. What's more, Huawei people had no background to support them. Without such resources, they were quickly in trouble.

Encircle the Cities by Occupying the Countryside

One of the older employees in the marketing department of Huawei recalls those days. In 1993, he heard that a particular customer was planning to procure equipment. Through the introduction of a friend, he secured the chance to pay a call on this customer. In order to catch his attention, the employee emphasized that Huawei's switching technology was on a par with equipment imported from overseas, and he implored the customer to come visit Huawei to examine things for himself. After his friend reinforced this invitation, the customer finally agreed. Once he got there, he was sadly disappointed. For someone who had been to see the facilities of Lucent, Ericsson, and other world-class companies, Huawei's production facilities and equipment were primitive in comparison. The examination went badly, and the two parted ways without consummating any business.

While the market in cities put up all kinds of barriers, the market in the countryside also presented Huawei with multiple setbacks.

Lucent, Ericsson, Siemens, and other international name-brand companies had made a deep impression on people already. Nobody in the world had ever heard of Huawei. This was an age in which fake and inferior goods were rampant in the market—customers simply did not believe that Chinese people could produce advanced telecom equipment. To prove themselves, salespeople hauled around slide projectors and samples. They traveled day and night and within one year had visited more than 500 counties nationwide. They came back with meager results, just a handful of orders. Trying to get business from a place called Yichun, on the border between Russia and China, a team of Huawei people including Zheng Baoyong and Li Yinan braved forty-below weather. These were high-level R&D people, braving the bitterest weather, and they did it dozens of times to carry out technical discussions, engineering reconnaissance, and equipment testing. They spent more than half a year making numerous long-distance treks to the border. On the day that bids were to be awarded, however, Huawei people were told that due to various reasons, the person in charge had left the department. At that point, a number of Huawei people who had worked on this project put their heads in their hands and simply cried.

In an internal meeting, recalling events from this period, Ren once exclaimed, "It was a great misfortune for Huawei to have chosen the field of telecommunications, out of the vast number of things we might have chosen. The misfortune lies in the fact that all industries are tough, when it comes to real business, but the field of electronic telecommunications is the toughest and riskiest. What's more, Huawei not only had no background on which it could rely, but also had no resources whatsoever. It was the fate of Huawei people to have to put out more sweat and tears than anyone else."

"Nevertheless," he concluded, "the market has no time for tears. It respects only the brave. If Huawei intends to survive, it has to carve out a bloody path for itself."

Each Failure Spawned the Seeds of Success
The moment it was founded Huawei ran into international competition at its very doorstep. Its opponents were global authorities in the field, millions of times more powerful than Huawei. Huawei could not cross swords with them directly; it faced the threat of being dismembered at any moment. This situation forced Ren to move his team's efforts to the rural market, where they could hunker down, unnoticed by the global giants in telecom. There he began to do a comparative analysis of Huawei's advantages and disadvantages with respect to its opponents. He settled on a sequence of very targeted war plans for how to engage in battle, which then drew the curtain open on the stage known in Huawei as "encircling the cities by occupying the countryside."

Overtaking the Technology of Shanghai Bell

Shanghai Bell[1] was an outstanding company. It was set up in 1984 as a joint venture that included the headquarters of the China Postal Industry Company, under the jurisdiction of the China Post and Telecommunications Department, the Belgian Fund for Development, and the company called Belgian Bell.

---

[1]Prior to the start of reforms in China, the only telecom entity was the Ministry of Posts and Telecommunications (MPT). To build a new system once reform began in China, and Alcatel [31.65 percent]) and the Belgian government (8.35 percent), and the MPT's equipment division, called the Posts and Telecommunications Industry Corporation. This latter partner retained a 60 percent share. The resulting entity was known as Shanghai Bell.

Belgian Bell was the European wing of the Research Institute of Bell, originally under the banner of America's AT&T. It developed the S1240 program-controlled numerical switch which was then in competition with AT&T's #5 model. Once Alcatel purchased it for a high price, Alcatel then assumed a 30 percent share in Shanghai Bell.

The buildup of China's telecom infrastructure gradually reached a peak in the early 1990s. Since the number of models under the "seven countries and eight systems" led to extreme chaos, and pricing was astronomical, Shanghai Bell suddenly began to get hot since its control shares were owned by China. For a while, all the heads of telecom bureaus in China hurried to Shanghai to visit the company. They lined up to purchase program-controlled switches. Indeed, within the short space of three years, Shanghai Bell's S1240 distributed control system took over more than 50 percent of market share of China's landline telephone market.

In 1997, new technology began to appear, with the V5 interface and access network, and Shanghai Bell found that it could not meet the demand. At this point, a precious pivotal opportunity appeared, when old models began to transition to the new technology. This had not drawn sufficient attention from decision-making levels, but Huawei was able to gather in the prize. It concentrated all its human resources and financial resources on developing network access products. It became a powerful weapon in contending against Shanghai Bell.

Sichuan can serve as an example. In 1998, Shanghai Bell held 90 percent of the market in Sichuan province. When Huawei came into Sichuan, it voluntarily provided network access products for free to such cities as Chengdu and Chongqing. After that, it gradually extended its activities to the sixteen other prefecture-level cities in Sichuan, as well as the three autonomous areas in the province. Once its networked operations reached a certain scale, Huawei "took advantage of the steel's being hot to beat on it." It extended its advantage from the network access gradually into program-controlled switches. Huawei then replicated the same model in other provinces across the country. It quickly came up to parity with Shanghai Bell in its newly added markets.

In 1999, the fixed-line telephone market gradually became saturated after years of ongoing investment in telecommunications. Huawei followed up its success in network access equipment with optical fiber networks, data transmission, and smart network products. As the fixed-line telephone business gradually shrank and broadband became the mainstream of business operations, Huawei's growth was given such a push that it was now hard to hold the company back.[2]

In 1998, Huawei ranked as number 10 among China's top 100 electronics companies, with its RMB 8.9 billion in revenues. For the first time, it overtook Shanghai Bell in the ranking. In 1999, with a clear advantage in technology, Huawei again ranked number 10 but now with RMB 12 billion in revenues. It consolidated its lead over Shanghai Bell.

Standing up to Lucent as an Equal

In early 1996, after America's AT&T and the Bell Laboratories were dismantled, AT&T became responsible for multinational telecom operations. Bell Laboratories, mainly an R&D operation, set up Lucent Technology. It also set up a production base in Qingdao whose influence radiated throughout China and the Asian region. Unlike in the years of the seven countries and eight systems, and different also from the other European and American telecom giants, Lucent not only had first-rate technology in global telecom but it also now had very strong customer relations in China since its equipment had come into China so early. It had superlative product quality as well as broadly based applications. As a result, once Lucent set up officially in Qingdao, in June 1996, it quickly signed contracts with a number of the telecom operating entities in China's more than 30 provinces. Lucent was the single most powerful competitor to Huawei in the China market at that time.

Given this fact, Huawei then also set up offices in three separate places in Shandong province, namely, Jinan, Yantai, and Qingdao. Once it formed a plan for full product solutions, it went into fierce competition with Lucent. It matched Lucent's efforts measure for measure.

---

[2]In 1998, Wu Jichuan was named minister of a newly formed Ministry of Information Industry (MII), with 34,000 people under him. This megaministry absorbed the old MPT and Ministry of Electronics Industry as well as major parts of the Ministry of Radio, Film, and Television. MII then became responsible for deciding on licenses for the new technologies.

In terms of business considerations, Lucent's control process switches were generally priced at USD 300 per line. Huawei set its list price at two-third that figure. In post-sales servicing and quality guarantees, Huawei took the initiative in providing customers with three to five years of free protection. Due to its clear price advantage and excellent service, Huawei quickly won broad customer approval.

Between the second half of 1996 and the end of 1998, Shandong was gradually linked up in a telecom network. First to be connected by rural fixed landlines were the thirty-one county-level cities in the jurisdiction that included the three cities of Jinan, Qingdao, and Yantai. After that, seventeen prefecture-level cities were gradually linked up in rural areas throughout the province.

Huawei grew with this buildup of Shandong's infrastructure, then copied the same successful strategy in the more than thirty other offices it had around the country. In the midst of fierce market competition, its telecom equipment "emerged from its husk," as it were, given the clear advantage of its business terms. Huawei quickly became customers' first choice. It began to rival Lucent as an equal in China's mainstream telecom markets.

After years of bloody and decisive battles, the wing forces of Huawei now came to the final stage of what is called *storming the fortifications* in Chinese strategy. That is, after encircling the cities by occupying the countryside, Huawei began to attack the cities.

In mid-1998, Ren decided to set up 270 more prefecture-level operating departments, using as their base the offices Huawei had already set up in China's thirty-two provinces, municipalities, and autonomous regions. Huawei now began to pay very high salaries as well as compensation in shares in order to attract high-level talent from universities and society at large. It recruited more than 10,000 professionals in telecom, computing, and sales. After rigorous training, Ren set forth a plan with respect to organizing personnel into teams. This plan presumed that each team was "cornered" and had to fight its way out. In each team, twenty-some people were selected from the managers of three departments—customer relations, products, and after-sales technicians. They were to fight for each inch of ground, throughout the 270-plus operating departments that Huawei now had nationwide. They were to regard their "unity of will" as an impregnable stronghold. Huawei may not have had product quality equal to that of multinationals, but it did have a "sea of people" strategy. By using its sea of people, the company had an absolute advantage when it came to human relations. It quickly took the initiative in owning the market.

For example, in Heilongjiang, Alcatel and other telecom giants had only four or five people handling the entire municipal as well as provincial market. Huawei had 220 people in its headquarter offices in that province. These were spread out over twelve operating departments, and even the most remote customer, such as an agricultural bureau, had someone specifically looking after him.

To highlight and strengthen its own advantage, Huawei tried to have an absolute advantage when bidding on tenders. It competed not only on price but also on total solutions. It candidly told customers what hidden costs might exist when using Huawei's equipment, not just the up-front costs, so that customers would know exactly what return on investment they were getting for the project, how much money they would make, and how much they would save. All this was clarified. During the "warring states period" of seven countries and eight systems, Chinese customers had been monopolized for years by European and American telecom giants. They now looked upon Huawei as a professional equipment provider but also as a partner who was trustworthy, someone with whom they could cooperate.

One thing was of particular importance. By cementing close relationships with customers through all levels of administration, Huawei collected and made use of large amounts of customer demand information. After sifting through this and analyzing it, Huawei got an accurate handle on the development trends of the telecom market. It then led the way in investing in the development of digital telecommunications, multimedia, and other new types of products. When broadband became the mainstream business in the telecom market in 1999, Huawei was well prepared. It began to wave the banner of "a broadband metropolitan-area network," and it mobilized its forces to move into this market.

This whole process can be regarded as an overturning of the monopoly that the seven countries and eight systems previously held over China's telecom market. Huawei's technical functionality now approached that of multinationals, while its pricing and servicing were far superior. In just four years, Huawei also completed a technological transformation. It went from being solely a fixed landline supplier to becoming a multimedia equipment supplier, and it realized the strategic shift from targeting the surrounding countryside to targeting the market in cities.

Universalizing Customer Relations

Once there is little difference in technology or products, the key to whether one wins or loses a customer becomes customer relations. Without exaggeration, it can be said that Huawei took over markets from competitors on this basis, which provided it with bamboo-like invincibility. Within Huawei, the strategy was even known as "using third-rate products to reach a first-rate market." The decisive element was Huawei's ability to create deep-rooted relationships with customers.

Ren commented at this point, "Huawei's products are not the very best, but who cares? If the customer chooses me over you, this is what one could call core competitiveness." Some telecom giants from Europe and America now began to fade from the picture. They could not seem to find operators to partner with them, and in a sense this was because they were their own worst enemies. In the strictest sense, rather than being defeated by Huawei, they defeated themselves.

For example, with respect to customer relations, multinationals commonly relied on a monopoly of their core technology. In dealings with the Ministry of Telecommunications and Posts, they focused exclusively on the decision-making level. Instead, Ren told his staff not to differentiate between large and small customers, and not to think of the positions of these customers as being high or low. All should be given the same importance. Staff had to stick close to customers at each level, and set up excellent relations at each level, as well as at each stage in the process from product promotion, to tendering bids on projects, to ultimately signing the contract.

Huawei was not in fact born with this knowledge. Its ability to set up iron-clad relationships with customers was also not achieved in one fell swoop, but rather came through experience and training. Its lessons were learned at the cost of bitter experience. Its success came from summarizing these lessons and then consolidating them into a system.

The market in Shandong can serve as an example. In 1999, Huawei's sales came to RMB 12 billion, to which the Shandong office contributed RMB 1.5 billion. However, within the seventeen prefecture-level cities in Shandong province, one particular market had zero sales, namely, Liaocheng City. The explanation was that the customer manager there treated that market in the same way as western companies handled their markets. He focused on the relationship with the decision-making tier and neglected customer relationships at other levels. Although he did sign a few contracts, customers were later unwilling to implement them because of problems with the various subsidiary departments in the telecom bureau.

Once a new person took over, he spoke frankly about the problems and the Shandong office of Huawei dispatched Jiang Chongbin to Liaocheng. Faced with a difficult situation, he did not really do any formal relationship building. Instead, he simply went to work together with employees as part of the team, pouring water, sweeping up, organizing the tables. After work, he would help customers with their children if they needed help on homework, or he would accompany a sick family member to the hospital, or help a retired employee who needed something. After six months of hard work, the Liaocheng Telecom Bureau evoked universally good feelings, from top to bottom. By having this contact at a close distance, they discovered that Jiang was honest and straightforward. He was a partner with whom they felt it was worth doing business. Because of this, he slowly won the trust of all levels of customers. Starting with a few tens of thousands of RMB worth of contracts, he won business. In 2000, at the annual awards meeting, the Liaocheng leaders even told its staff to "learn from Jiang Chongbin's diligence and willing spirit." In that same year, Shandong became the champion sales earner in Huawei. The Shandong office took in RMB 2.2 billion in sales; of that amount, contracts from the Liaocheng Telecom Bureau contributed close to RMB 200 million.

Similar examples of turning defeat into victory later became common in Huawei's other markets.

Ren reinforced this approach at the 2001 awards ceremony of Huawei. Once he had listened to the reports of managers from thirty-some offices around the country, he said gravely,

Customers are the ones who clothe and feed us. They are like our parents. You absolutely must not look on the customer relationship as being purely a business relationship. You certainly do not want to just chase after the boss to sign contracts, and disregard people on down the line. Not only is this a principle in how to be a human being, but it is a principle to live by if one wants to gain a market.

Focusing on universal customer relationships is not just something for the marketing department. It goes for the entire company. This means that you are glad to meet with anyone. You must not fail to provide support just because the person you are dealing with is a minor engineer, and you therefore don't introduce products to him. We want to stick close to the customer at each and every level, each rank. We want to share in all our customers' concerns. If we do, customers will give us a ticket. Here a ticket, there a ticket. Added up, they come to a lot of tickets.

Each and every one of Huawei's 110 sales departments around the world then began to implement the principle of *universally applied customer relations*. The practice permeated all parts of the company and has continued to this day.

Always Feel Grateful

In fact, of course, universally applied customer relations is not something that belongs to Huawei alone. To open up markets, all companies commiserate with customers in hard times and congratulate them in good times. Maintaining excellent relations with every tier of customer is commonplace. However, once a given piece of business is over, or the decision maker on the customer side is transferred elsewhere or retires and loses the power to make decisions, it is rare for a company to maintain a feeling of gratitude toward him and a friendly relationship. This is where Huawei is different. It is key to the emphasis that Huawei puts on adhering to the principle of universal customer relations.

For example, in the early days of Huawei, older experts in China's post and telecommunications system provided Huawei with enormous support. When they were pulled back to the "rear lines," Ren gave special instructions to the sales department:

> You are to pay periodic visits to any older experts who contributed to Huawei in any way. What's more, you should invite them to visit Huawei's headquarters in Shenzhen, just to look around. In addition, you send every telecom employee who is now retired but who helped Huawei in the past a small remembrance on his birthday or at New Year's, some fresh flowers, or a birthday cake. He will be glad for it, but his leader will also be glad for it. Always maintain a feeling of gratitude for the help that customers gave us in the past. Huawei not only needs to have a good reputation around the world, but it needs to have a positive influence around the world, one that comes from the behavior of our Huawei employees.

In 2000, China's postal administration was separated from telecom. Authority to procure equipment was withdrawn from county and municipal-level bureaus to the provincial level. Because of this, some senior officials wrote Ren, recommending that he too reduce the numbers of employees in his offices. They even recommended that he get rid of the more than 270 regional operating departments across the country, to lower costs. This suggestion was vehemently rejected and indeed criticized by Ren.

As a result, Huawei did not eliminate the regional operating departments, but it continued, as before, to maintain excellent relationships. Events were later to prove that although equipment is procured on a centralized basis at the provincial level, the choice of what types of equipment to buy is still made at the provincial and municipal levels. These actual users are the ones who make direct requests for equipment needs and who recommend manufacturers. Huawei had been close to customers all along, at all levels. Each time the provincial bureau invited tenders, Huawei naturally became customers' first choice.

History is not comprised of a series of startling major events. Instead, it has many small stories behind it, tiny details that propel it forward. Inside the company, Ren often reiterated this lesson to employees: "Huawei has no big things. We create big things through the accumulation of small things. If you do common things well, they become uncommon. If you do simple things well, that is not simple."

In fact, greatness is also composed of the commonplace. Huawei's success was not due to any earthshaking major circumstance. Instead, it was created in the most common and unsurprising ways. Since Huawei people maintained this attitude of gratefulness, however, and since they were honest with both new and old customers, for over twenty-seven years they were able to paint a scroll of remarkable achievement.

Ultimate Honesty and Trustworthiness

In June 1999, at the very first graduation ceremony for the training class of new employees of the marketing department, Ren sat together with students and asked, "What is the highest realm of good faith and integrity?"

One student replied without thinking, "You do what you say you'll do." Ren said that this was the initial stage of honesty. Another student thought a bit and then said, "The highest realm of honesty and trust is to safeguard the quality of products." Ren said that this was a basic precondition. Everyone fell silent after that. They shook their heads and said they didn't know. Ren said, "The highest realm of good faith and integrity is that the company does well, does not go down, and thereby protects the customer's investment. That is its greatest value to the customer. What that means is we have to grow together with customers, develop with them. Making customers succeed is the same as making ourselves succeed."

Anyone who has worked alongside Ren has a deep understanding of his willingness to invest in customers. His private life is simple and frugal. He still uses the old teapot he used for years in the army in his office, and he has a simple military cot set up there. The moment you start to talk about customers, however, and how Huawei needs to invest in the basic infrastructure that helps customers, he stops thinking about costs. In 2001, even as Huawei's "winter" was approaching, Ren explicitly addressed this issue. "The one thing under heaven that gives Huawei money is the customer. If we do not serve the customer, who will we have left to serve?! Since customers are the ones who determine the life or death of a company, customers are our sole reason for existing!" This can be regarded as the key element in the market strategy that Huawei has consistently pursued.

In 1995, during the seven countries and eight systems period, since equipment came from seven different countries, models were chaotic and it was hard to coordinate capacity. This led to the phenomenon of rampant fees and granting of kickbacks, and it also led to frequent paralysis of systems. Since causes were complex, each manufacturer could blame the others. Matters were either delayed or never resolved, and operators were then exasperated beyond endurance.

In that same year, Huawei's financial situation turned slightly better. Ren therefore invested RMB 100 million in an office building on Science and Technology Road (Kefa Road) in Shenzhen's Tech Industrial Park. He personally named this building the Huawei Customer Service Center Building. At the same time, he set up thirty-two customer service centers in China's thirty-two provinces, municipalities, and autonomous regions, which operated in tandem with the local Huawei offices. To lessen the chaos in the nationwide telecom network in China, Huawei then announced that any customer whatsoever could come in for assistance. It did not matter whose equipment was at fault. Ren told his staff that Huawei's customer service centers had to respond as fast as possible and go on-site if necessary to resolve any problem.

As an example, in the spring of 1996, the Telecom Bureau in Zhangjiakou was suddenly paralyzed and failed to function. It used overseas equipment. Since this occurred over the Spring Festival holiday, technical personnel were all home on vacation. Given the emergency, this customer asked for assistance from Huawei's customer service center in Hebei. Huawei's technical personnel immediately rushed to Zhangjiakou to provide urgent assistance, braving a storm in the middle of the night to get there. Phone service was quickly restored to normal. The customer was extremely moved by this and developed a strong friendship with Huawei. Later, the provincial bureau of Heibei also gradually began to substitute in Huawei equipment for that of other countries.

In 1998, Huawei's revenues reached RMB 8.9 billion. This coincided with a time when real estate prices were soaring nationwide. After ten hard years, Huawei now had plenty of capital but Ren did not, like many other well-known companies, take the company into major investment in real estate. He also did not list the company on the market, nor did he diversify. Instead, he invested RMB 6 billion to buy a piece of land 1.3 square kilometers in the Long-gang district of Shenzhen, at Bantian. There he built the world's largest production base for telecom equipment. The investment just in the hardware and software for Huawei University cost RMB 1 billion. This cluster of buildings, built to train both customers and employees, occupies 275,000 square meters and is composed of living quarters and teaching areas. Some 85 percent of the total area is covered by greenery. The teaching area has extensive computer facilities as well as more than 200 classrooms. It can simultaneously accommodate more than 5,000 customers at a time as they undergo training.

Huawei University proved its worth after the global Internet bubble burst. In 2003, many customers who had studied there now regarded Huawei as substantially different from multinational companies. They lined up to sign orders with Huawei for considerable sums of money. By the end of 2012, Huawei University had provided several million person-times of professional training for operating entities and corporate customers, and not just from China but from more than 100 countries around the world. Not only did this elevate Huawei in terms of trust-worthiness in the minds of many customers, but also it trained countless numbers of potential customers.

"Making customers succeed is the same as making ourselves succeed." This phrase of Ren's became the compass for Huawei as it "stormed the fortifications" in going after markets. Results were substantial. Although the massive amount of investment reduced Huawei's profits in the short run, it helped create much greater long-term value. The process illustrated Huawei's basic principle of how to establish a foothold and grow.

CHAPTER THREE

THE SOUL OF INDEPENDENT R&D

Innovation may be hard, but it is the only path to take
if an enterprise hopes to survive. In addition to just survival,
it is the path to success.

—REN ZHENGFEI

In a meeting within the company, Ren once described how China had fallen behind in the course of the past century. Due to historic reasons, it was engaged in warfare, civil war, and closed-door policies as the second industrial revolution was going on elsewhere. Production methods that were both highly refined and completely western were lifting quality and lowering costs in other countries, and keeping products in the forefront over the long run, thereby occupying global markets. Only in 1978 did China open its doors a crack to take a look at what was happening. It observed that the world's economic patterns had undergone revolutionary change.

Japan serves as an example. During the revolution that occurred in the electronics industry in the 1970s and 1980s, the wealth that was generated by Japan's enterprises, which emerged as a group, was quite enough to buy out the entire U.S. stock market. In the information age of the 1990s, however, under the impetus of venture capital and innovations, thousands of American companies invested crazy amounts of money in R&D. This generated an extremely large number of patents, inventions, and core technology. It had the effect of propelling the high-speed growth of the information age and of solidifying American prosperity by holding 60 percent of the global information market. When IBM was at its height, the market value of its shares surpassed the combined value of the German stock market. IBM became the single most profitable company in the history of global economic development.

At the end of 1997, when Ren traveled to the United States on an investigative tour, he discovered that IBM's asset value that year was USD 75 billion. In contrast, the asset value of China's entire electronics industry, after 20 years of reform and opening up, was USD 15 billion. It was a mere one-fifth of IBM's. Meanwhile, as the "dragon head" number 1 company in China's telecom manufacturing industry, Huawei was putting RMB 400 million into investment in R&D in 1997 while IBM was putting in USD 6 billion. The average annual investment in R&D of America's IT companies, including Intel, Hewlett Packard, Lucent, and Cisco, came to somewhere between several hundred million and several billion dollars.

The gap may have been caused by historic reasons, but nonetheless it was a fact. Faced with it, Huawei was going to have to rely on more than some kind of spirit if it intended to catch up and take the lead. It had to change its modus operandi, realize that its poverty was forcing it to generate change. It had to find an effective path by which to shorten the distance between China and world-class enterprises, and this path had to be paved with mechanisms that enabled innovation.

The Principle of Concentrated Pressure

Ren often reminds his R&D staff that a nail can be very small and still go through hard material. The key is to apply concentrated force in just the right place, so that a tiny sharp point focuses the pressure. As described above, in 1993, when Huawei's financial resources were extremely limited, Ren focused such money as he had on developing the C&C08 digital switching system. As a result, he succeeded in breaking through with this core technology and achieved massive success in the market.

In 1996, six professors at China's Renmin University began to draft what became known as the *Huawei Basic Law*. This used the principle of focusing pressure to describe how, by reasonable ways, to make technological breakthroughs using limited resources. It includes the following description. "In terms of successful key elements and selecting strategic development points, in order to get beyond your opponent's ability to assemble massive resources, either you go for a particular market ... or you don't. If you do, you focus all of your resources on it. That means concentrating human resources, material strength, and financial strength on one key point in order to achieve a breakthrough."

This can be regarded as the very essence of how Huawei uses the principle of concentrated pressure to achieve breakthroughs in core technology.

Despite its relative paucity of financial resources in 1993, Huawei chose to position its R&D on the C&C08 digital switching system. It did not choose a product that had a low barrier to entry, or that required little investment and would see quick results in the market, as the target of its attack. Instead, it bravely targeted the AT&T technology called the #5 switch, at that time the global leader. Later facts were to prove that since Huawei selected the "highest point," it avoided the risk of being washed out of the market as a result of having to live with outdated technology. At the same time, it increased the distance between itself and other domestic competitors. Huawei drew itself closer to the ranks of international competitors during two different warring states periods, one the period of seven countries and eight systems and the other the period of civil war. As a result, Huawei ensured that its products and technology were cutting-edge R&D. As soon as its products appeared, they set a high bar, which was critical to reducing and dispersing R&D risk.

In the course of breaking through the control process switch and moving toward higher technological spheres, Ren became aware that the overall strength of the company was insufficient. Instead of hyperactive behavior that tried to go in all directions, he went systematically from one product to the next. He took the profit that had been made from the control process switches and invested it in R&D on optical-fiber transmission. Once this product had been launched and approved by the market, he took the profit made from optical-fiber transmission and invested it in R&D on digital telecom and wireless products. In this cyclical manner, he kept from reaching for too many things at once and losing focus on what was important. After 27 years of growth, he gradually developed the most complete technological advantage in the entire sphere of global telecom.

This method of "transferring the baton" from one successful product to the next not only resolved the chronic financial scarcity suffered by all *minying* companies, but also allowed Huawei to make full use of what at the time were relatively weak R&D human resources. At the same time, this stable and methodical way of expanding technology also allowed the company to avoid having several fronts trying to move forward at the same time. It avoided the potential risk of incorrect policy choices.

One very critical aspect of all this was that Ren did not seek to diversify after gaining returns from each of a series of core technologies that Huawei put on the market. Instead, he made a hard-and-fast rule: Every year, he would invest at least 10 percent of sales revenue back into R&D. This was true even in 2000, when telecom markets shrank around the world. Even in that year, Ren did not reduce R&D investment. On the contrary, he increased it by 5 percent over the previous year, to the amount of RMB 3 billion. In the second half of 2008, as the global crisis was underway, competitors all busily cut their R&D investment, but Huawei's investment in R&D instead exceeded RMB 10 billion. In 2011, when the industry as a whole hit a ceiling, Huawei's R&D exceeded RMB 20 billion.

From the perspective of globalized technological competition, Huawei's massive investment in R&D has reached a point that no contender within China can match, or even dares to try. However, compared to the investment put in by IBM, Microsoft, Intel, Cisco, and other famous companies, it is small potatoes. However, Huawei's focus and ruling spirit nevertheless allow the company to be prepared for any threat. In 2003, Cisco used a strategy of charging infringement on intellectual property rights to attack Huawei. Both parties ultimately shook hands and came to terms. Afterward, in an interview with international media, the CEO of Cisco, John Chambers, declared, "In the future, Cisco will have only one opponent. That one comes from China and is Huawei."

In the 1990s, the use of *concentrated pressure* was also the kind of measure that burned bridges behind a company and made it go all out for one goal. Nowadays the situation may be different. This method of achieving breakthroughs in core technology is not the only way, given the more complete mechanisms that allow for funding inside China, including venture capital. Nevertheless, Chinese corporations would find it worthwhile to learn lessons from Huawei about this method of concentrated pressure, in terms of how to focus on not just the implementation of decisions but also the energy of a corporation.

"From the time it was founded, up to now, Huawei has in fact dedicated itself to doing just one thing. It has not been seduced into trying to go for other opportunities. It has steadfastly maintained its focus on researching core network technologies in the telecom field. In this, it is duty-bound not to turn back." This statement by Ren should be read as an important lesson since it illustrates how Huawei's principle of concentrated pressure could overturn the situation in world telecom.

The Purchase of Patents and Innovation

For Huawei, without its own core technology, taking this practical path of directly purchasing patents had one single goal, namely, to reduce the gap between itself and its competitors as fast as possible and to participate in international competition as soon as possible.

Optical-fiber transmission can serve as a good example. Huawei has a certain advantage in the technology that handles short-distance optical-fiber transmission. However, it does not have an advantage in long-distance transmission. In this arena, an American firm in Silicon Valley is the world leader. Starting in 1995, that firm began investing several hundreds of millions of dollars into R&D in this area. In 2002, when the IT bubble burst on a global basis, this optical-fiber telecom company declared bankruptcy. At the auction for the company, Huawei purchased the technology for the low price of USD 4 million. After secondary development that took nine months, Huawei successfully developed optical-fiber technology that could transmit over a distance of more than 4,600 kilometers without using electric cable. Once launched on the market, this technology was quickly applied on a widespread basis. In 2003, Huawei won the contract as the sole party building the world's longest optical-fiber transmission project, a Russian project that extended for 18,000 kilometers. In 2004, Huawei again won the contract as a sole party for the largest operator in Latin America, Telemar, for a large-capacity long-distance backbone optical-fiber transmission line, the first of its kind in the world. In just two years, from being an outsider in this field, Huawei advanced to being the industry's leading company.

In terms of international patents, Huawei has always upheld a principle that "respects other people's intellectual property rights while at the same time focuses on protecting one's own intellectual property rights." In line with international precedent, Huawei has paid for the use of other people's intellectual property rights through proper legal methods and through payment of required fees. In addition, it has signed a series of agreements that exchange intellectual property rights with the world's giants in telecom— Ericsson, Alcatel, Nokia, Siemens, Qualcomm, and others. For example, the broadband technology called DSLAM was invented by Alcatel, whereas the technologies associated with base stations, simple mode, were invented by Huawei. After two years of negotiations, Huawei ultimately came to a patent-exchange agreement with Alcatel on these, after which Huawei's market share in the DSLAM field shot up in a straight line. In 2007, the company ranked second on a global basis, and in 2008, it surpassed Alcatel and ranked first, with 31 percent of the market. What's more, this record has been maintained to this day.

Via procurement of patent rights, Huawei not only can obtain core technology in an indirect fashion, but also can participate in international competition on a faster basis.

Of particular importance is a significant move that Huawei made with respect to basic and advanced research that it was not well versed in itself. That is, Huawei accumulated core technologies by means of cooperation with others during China's reform and opening up process. For example, it jointly developed the optical-fiber network, CDMA,[1] and digital telecommunications technology with a number of institutes

---

[1]CDMA is the acronym of Code Division Multiple Access. GSM stands for Global System for Mobiles. These are acronyms for the two major radio systems used in cell phones. Most of the world uses GSM because Europe mandated the technology by law in 1987 and because it comes from an industry consortium. CDMA is owned by chipmaker Qualcomm. The main change from analog to digital occurred in 1995 and 1996, and many U.S. carriers went with CDMA because it was newer and more advanced. GSM came first but caught up; China used both. The future is LTE, an acronym for long-term evolution. In China, the license for GSM is held by China Unicom. The license for CDMA is held by China Telecom.

of higher education in China including Tsinghua University, Peking University, and the Beijing Post and Electronics University. In the sphere of applied technology, Huawei set up experimental labs and carried out joint development with some of the world's most famous companies, including IBM, Intel, Microsoft, Texas Instruments, Agere, ALTERA, and SUN. In the sphere of end-user and 3G and 4G technology, it has set up joint-venture companies and R&D centers with other famous companies including NEC, Matsushita, Siemens, and Motorola. Together, these facilities allow for risk mitigation and for profit sharing in the industry's chain of value. Through such broadly based cooperation, Huawei not only resolves difficult issues in its own basic research, but also nurtures high-end human resources on its own behalf. Such people are now working at international levels in R&D. They have played a critical role in propelling Huawei's takeoff in terms of its technology. When one looks at the evolution of high-tech in the information age, practice of the bring it in approach has been generally accepted in Europe and the United States. For example, Microsoft's DOS, Windows, and other core technologies were both influenced and inspired by advanced technologies in the world. Through consistent upgrading and innovations, the concepts were put into products that could satisfy market needs. Because of this, Microsoft achieved massive commercial success. In contrast, China lacks the kind of innovative atmosphere and commercial base that one finds in the United States. Nevertheless, in a short time, China has been able to reduce the enormous distance between itself and world-class competitors, and the bring it in approach has definitely played a role in this. It has enabled Huawei to pay less in tuition and take fewer unproductive detours along the way. From this perspective, the purchase of patents and innovation, Huawei's bring it in approach has brought in not just core technology but also the thing that China needed most if it was to catch up with Europe and America, namely, time.

Technology is Always Only Ahead by Half a Step

On April 28, 2005, in a forum about the direction of Huawei's research, Ren joked about how Huawei is a vegetarian. He said, "Some telecom companies in the west generate innovations at a faster pace than anyone. They are meat eaters, which is why they can run so fast. Huawei instead is a vegetarian. That's why we run slowly and won't be able to catch up with western companies. However, western companies also run too fast, too precipitously. They may fall off an overhanging cliff if they're not careful. Huawei runs carefully, and once its foot touches the edge of that cliff, it looks over and sees them down there. Once that happens, it quickly reins in the horses."

Despite its jocular tone, this vividly describes how the winner in the marathon of high-tech is not necessarily the front runner. Although super-advanced technology is a treasure for all humankind, we should not pay too great a price for it. Often, that price can include sacrificing ourselves.

Fujitsu and NEC can serve as examples. In the age of analog switches, these two Japanese companies had an absolute front-runner position, but when digital switches came along, they slowed down their pace of innovation and were overtaken by Lucent, Siemens, Alcatel, and other European and American companies. To reverse their losses, Fujitsu and NEC developed a switch that was more advanced than digital switches, namely, the 400G ATM switch. However, this switch was more advanced than the market was ready to absorb or could apply. It was three steps ahead of customer demand, and in the end the companies were washed out of this market. Similar things have also happened to Nortel, Motorola, and other telecom giants in North America.

As a company in the category of "national treasure" in America, Motorola is the original ancestor among wireless telecom companies. After inventing the world's first mobile telephone, Motorola had no competitors in the mobile telecom sphere for quite a long time. When GSM wireless technology arose in the 1990s, most notably with Nokia and Ericsson, Motorola, threatened by this, nevertheless failed to pay sufficient attention to the unique features of GSM technology and move ahead with the times. Instead, it invested USD 5 billion in developing a more advanced "Iridium plan." This plan contained too much in the way of technological bells and whistles, however. When the product hit the market, it was so advanced that customers had no idea how to use it. In 1996, the first models had excellent connectivity, but it cost several U.S. dollars to talk for just one minute. Using GSM cost just a few pennies.

Since Motorola had become distanced from real market needs, it gradually also moved further away from the mainstream navigation required by wireless telecom technology.

Although the Iridium plan did not make the grade, Motorola did not in any way give up its pursuit of new technology. Another company made the same mistake, however, which led directly to its bankruptcy. Nortel had produced the very first telephone in the world. It had some 100 hundred years of glorious history behind it and a wealth of technology, and it was a giant in the field of North American telecom. After the IT bubble burst in 2000, operators that had been hit the worst no longer pursued high-tech in the same blind fashion. Most players in the industry tried to think of ways to optimize networks via the Internet Protocol (IP), in order to save on investment. In contrast, Nortel put large sums of investment into researching the next-generation network (NGN). Moreover, it missed the first opportunity as systems changed. It perceived the shift in the market too slowly. In 2004, as 3G was gradually becoming hot around the globe, Ericsson, Alcatel, Huawei, Zhongxing, and other telecom equipment makers quickly either upgraded their 2G models to 3G or extended and upgraded them. Nortel yet again mistakenly sold its 3G technology to Alcatel and then took the money and invested it in what was the most advanced LTE at the time (long-term evolution, colloquially called 4G technology). Seen from today's perspective, Nortel was indeed the one to push the wave in terms of new technology. Unfortunately, it also became too far removed from customer needs. It was too technology oriented when systems changed repeatedly. After the financial crisis of 2008, Nortel lost hope. It had been mired in losses for eight years. Its creditors sold its core technologies at a discount to Microsoft, Apple, Google, and Ericsson, and it disappeared altogether from anyone's field of vision.

Ren had the following to say about the great companies that in the past made such enormous contributions to humanity. "These tragic companies failed not because their technology was not advanced enough, but because it was so advanced that people did not understand it and therefore did not adopt it. They then lost their competitiveness. However, it is almost impossible for an enterprise not to have advanced technology. Huawei's view on this is that we must maintain a lead in terms of innovations but that lead should be just half a step in front of our competition. If we are three steps ahead, we become a martyr."

To strengthen the awareness of R&D staff about market considerations and avoid being trampled, Ren strictly requires that 5 percent of the R&D staff switch jobs to marketing every year. At a banquet in which he was sending R&D staff to their new jobs in "frontline" marketing, Ren made this analogy. "Technology people should not worship technology as though it were some kind of religion. They need to be process-engaged businessmen. Your technology is used to make money. Technology is not worth anything until it is sold. If Huawei stubbornly holds onto the idea that it absolutely must have the world's most advanced products, Huawei will starve to death. We will become dead in a vase, like Van Gogh's sunflowers."

He added, "Technology can become as advanced as it likes, but if it cannot create commercial value, not only is it not significant to the company but it can bring on massive losses."

To strengthen this concept of linking practice to the summation of actual experience, of carrying on R&D in ways that tread on firm ground and meet real needs, in the second half of 1998, Ren declared the following during an awards ceremony for Huawei's R&D division:

> If anyone wants to talk to me about Huawei's strategy as a company, I'm not interested. Why? Right now, Huawei's problems are not strategic. Our problems instead relate to how we are going to continue to survive.
>
> Everyone sitting here is quite young. You are all sunflowers, looking toward the future. But the greatest problem of youth is that it is inexperienced. This company has developed very fast. Not only do you not have a basis in theory but you also have no basis in practice. If we encourage everyone to "raise ideas, have strategic plans," in my view, Huawei will become no more than a thin reed. When the wind blows ever so slightly, it will be blown over.

Under the guidance of this kind of thinking, Huawei has gradually set up an R&D policy line that takes customer satisfaction as its guiding principle. A number of things have been derived from this, such as award mechanisms and "materially-based spirit, small advances, large awards, no encouragement for great big ideas." These have gradually evolved into being a corporate culture that is unique to Huawei. This came about particularly after telecom went through the baptism of fire during the 2001–2003 global "winter freeze" in telecom, when Huawei people saw with their own eyes how giants in telecom could go under despite super-advanced technology. This confirmed their belief that customer needs should guide R&D. It led to the establishment of rules that would allow the company to continue to exist on a long-term basis.

For example, in 2004, when it was breaking into the European market, Huawei took note of the fact that Europe is fairly crowded and therefore pays strict attention to environmental protection. The company promoted a 3G distributed type of base station which was only as big as a DVD box. It could be installed inside the building or hung outdoors, it saved electricity, and it was environmentally friendly so had a "green" image in customers' eyes. This is Huawei's "small advances, large awards, no encouragement for great big ideas" typical case.

There are many similar cases in which Huawei was able to be successful. Although each was unique in some aspects, their commonality was that the functions of the product were extremely practical. They satisfied customer needs to the ultimate degree. This was also a key element in why telecom giants going after super-advanced technology were gradually withering, while Huawei was making huge strides forward. As a Harvard Business School professor has revealed, "When breakthrough technologies appear, they can change the entire rules of the game in the industry. However, these so-called breakthrough technologies are often not in fact the most advanced technology. Instead, they are the ones that stick closest to the market."

Not Taking Risks Is in Fact the Biggest Risk

On December 21, 1995, Prof. Wu Jiangxing came to visit Huawei. He was the highly eminent person who had developed China's first 04 program-controlled switch, and he was also CEO of the telecom company Julong.

Although Ren was an important competitor, Prof. Jiangxing felt he had to warn Ren when he learned that Huawei stuck rigidly to the rule about investing at least 10 percent of its sales revenues in R&D every year. He was deeply familiar with the risks of overinvestment in R&D. "When Huawei invests such a huge amount in R&D, how does it avoid major risk?"

Ren responded decisively. "Research is indeed a kind of gambling. If you are not in the game, however, if you don't dare to be in the game, you're left behind. And that definitely dooms you to failure!" In looking back at the major development of China's telecom markets in the 1990s, it is clear that many companies grabbed hold of historic opportunities at the time. In addition to Julong, Datang, Zhongxing, and Huawei, the country had some 200 other outstanding corporations in the business. Given sustained and intense market competition, however, some took a fateful path in the wrong direction because they had different ideas about how much to invest in R&D.

In 1996, when the world was immersed in the very hottest time for control-process switches, Huawei became aware that the market for fixed-line telephones would become saturated. It therefore put large amounts of R&D investment into the IP arena. In 1997, when network-accessing products became the hot spot in the market, people around the world generally thought that they were already secondhand and could not be developed much further. People adopted the attitude of "planting their feet in one place to wait and see what would happen." Instead, Huawei and Zhongxing took the risk of burning bridges behind them and going all out in this direction. They swiftly put out accessing products and were immensely successful. When other companies finally woke up and began to take action, these two were already monopolizing the market. After making enormous profit from this technology, Huawei and Zhongxing went on to make breakthroughs in core technology including optical-fiber transmission, digital telecom, GSM, and 3G, among others.

In 2006, as Huawei and Zhongxing vied for being in the ranks of the top 10 emerging high-tech companies among global telecom equipment manufacturers, other companies were out of the running despite extremely strong R&D capacity. When the two main opportunities appeared, these companies were too slow by half a step. They could not catch and ride the huge waves of an information age that was changing by the day. In the end, they also could not return to their own mainstream business. There were many reasons for this, since these were extraordinarily creative, outstanding companies. It is unfortunate that they could not display their brilliance on the international stage together with China itself.

As we look back on this stretch of history, an information age when things can change in a second, we see that the key to a corporation's future can lie in just a few important steps. It can even lie in the orientation of just one product. However, when opportunities arise, R&D risks can also be terrifying. In this regard, Ren had unusual intuition.

In a knowledge-economy age, the methods by which a company survives and grows have also changed in fundamental ways. In the past, one relied on doing things accurately. Now, it is more important to do the right [accurate] things. In the past, people thought that innovations were taking a risk. Now, if you don't innovate, you are taking a major risk.

Thinking back, if we did not take risks and simply followed along behind people, staying in second or third place for a long time, we would not be in any position to compete with multinationals. We therefore would not have the right to stay alive. If we had kept to the old ways out of habit, we also would not have been able to develop at the speed we have.

Ren clearly has a risk-taking spirit, but his willingness to do something includes reading opportunities accurately. This honed kind of entrepreneurial courage can be regarded as another key factor behind the emergence of Huawei on the world's high-tech stage.

Do What You Are Best at Doing

In the sphere of high-tech, whoever holds core technologies in his hand is the one who holds the strategic high ground when it comes to occupying a market. As swordsmen who actually occupy this strategic high ground know, however, what they ultimately are judged on is not the sword of core technology, but rather how they wield that sword. The one who holds a double-bladed sword, in terms of both low costs and core technology, is the one who makes himself invincible.

The reason Huawei has been able to defeat so many strong competitors in international markets is that it has fully wielded China's innate advantage in having low-cost R&D.

In the sphere of telecom, the key technologies were invented in other countries. That goes for process-control switches, broadband, 3G technology, and others—all were invented elsewhere. In this regard, Huawei has adopted the policy of being a technology follower. Prior to Huawei's entrance into the telecom arena, overseas companies had invested more than 10 years, indeed 100 years in this field. They had amassed a large quantity of research results and patented technology. In contrast, China's creativity had been enormously repressed due to historic reasons. It has been impossible to even begin to think of comparing China to the West in terms of its atmosphere for entrepreneurship or its mechanisms for enabling business. It would be very hard, therefore, to think of inventing such core technology as Microsoft's Windows on Chinese soil, not to mention Intel's CPU, or Cisco's routers. Even creating patented technology for similar things takes enormous input into R&D, and it also requires at least a decade to be commercially successful. Huawei had absolutely no choice but to be a follower.

Despite being in an inferior position when it came to core technology, however, Huawei found a way to realize its own advantages and carve out its own path. These advantages came in the form of making noncore special-use chips. Since Huawei had a vast quantity of low-cost R&D personnel, it could do best at carrying on secondary development or deep-seated development outside noncore areas of technology. In that way, it could construct a certain degree of technical advantage and then expand and extend that via its cost advantage.

As early as the fall of 1992, when Ren took his first investigative trip to the United States, he was already aware of the field in which Chinese people could most excel. The key chips that millions of control-process switches require cost USD 200 per chip. If you are buying them in quantity, you still only reduce the price by a few dollars. If you can research and design your own chips, then take them to Silicon Valley in the United States, or Korea, or China's Taiwan for final processing and production, the cost of each chip comes to only around USD 10. If you do that in wholesale quantities, you can reduce it by 50 cents to a dollar. The difference between your price and the original price is now 20-fold. Given that control-process switches are composed of several hundred large and small chips, and given that you have cut your cost of the key chips by USD 30 to 100 each, you are saving a very substantial amount of money. Clearly, if Huawei could make breakthrough advances in the realm of chips and also take advantage of China's innate advantage in low-cost labor, it could wield an enormous cost advantage in the market. It could definitely emerge as a contender among a field of competitors.

To that end, in 1992, Huawei organized an "integrated-circuit design center" that was responsible for doing R&D on silicon chips. After four years of hard work, in July 1996 this center finally successfully created the unique chip technology used in the 100,000-*line* grade and 1 million-*line* grade switches. After absorbing this experience, Huawei began asserting its authority in the successive fields of telecommunications, broadband, multimedia, GSM, CDMA, and 3G. It wielded its advantage like a sharp knife splitting through bamboo.

In 1998, the chip project alone saved Huawei a large amount in procurement costs as the company realized RMB 8.9 billion in revenues. The "deep" development of chips became critical to Huawei's reduction in costs and resulting competitive advantage.

In 2003, Huawei successively developed and produced the WCDMA basic station set of chips and the number one 10,000-*line* class of chips, which showed that its chip design was already reaching international levels. In October 2004, Huawei spun off its integrated-circuit design center and made it independent. It set up a company called the Hai-Si Semi-Transistor Company, aimed at global chip design and process production. As Huawei's internationalization continued and as the company became stronger, this Hai-Si Company swiftly grew in size as it set up branch institutions at Huawei's 17 R&D centers around the world. This not only concentrated world-class R&D personnel, but also enabled researchers to take advantage of international design standards and testing and verifying technology. After years of unstinting hard work,

Hai Si has become the "new aristocracy" in the global arena of chip design. It has more than 2,000 self-generated R&D special-use chips that are at an international level in terms of technology and that are internationally patented. According to China's semiconductor industry association, by 2011, Hai Si ranked first among China's top semiconductor companies. With assets valued at RMB 6.67 billion, it is the indisputable "chip giant" in China.

Using a Management Advantage to Stimulate a Cost Advantage

Although Huawei possessed the human resource advantage of low-cost labor in China as it developed, its R&D results were poor due to poor management. It therefore wasted a tremendous amount of resources. Meanwhile, other Chinese companies such as Zhongxing were also emerging at the same time, which led to intensified price wars. To lower their own costs, multinationals also started shifting their R&D centers to India and to China. Huawei wanted not only to internationalize itself, but also to move its R&D institutions overseas. This caused it to lose its original cost advantage as costs gradually approached those of the giant telecom firms in Europe and America. The most important consideration for Ren now became how to improve management in order to increase efficiencies and lower R&D costs.

In 1998, Huawei invested RMB 2 billion to bring in IBM's R&D management system known as IPD, which is an acronym for integrated product development. Under the guidance and assistance of IBM consultants, Huawei strode briskly into a period of restructuring. The intent was to go from being technology-driven to being market-driven. IBM consultants were amazed to discover, for example, that Huawei developed the entire spectrum of its products in-house, including the most common integrated circuits and noncore software. This inevitably led to low efficiencies and affected the time it took to get products to market. IBM vigorously recommended that Huawei develop its own core technology, but use the front-runner advantage of the rest of the world to supplement its own deficiencies for things in which it did not excel or for which it had outmoded technology. When establishing R&D priorities,

Huawei now began to diagnose its own advantages and disadvantages more accurately. It then began to optimize its R&D processes, deciding what to do itself and what to bring in from elsewhere, via purchase of patents. It had only one goal, which was to satisfy market demand as quickly as possible and make a profit in doing so. After 10 years of paying respects to its teachers, Huawei was able to ensure that IPD concepts were infused into the very bloodstream of all Huawei people. Not only did this vastly reduce the time cycle of R&D, but also it improved results and lowered management costs considerably.

Another great change in Huawei's R&D systems came with Capability Maturity Model for Software (CMM) management. As a hard rule, Huawei takes 10 percent of revenues every year, in capital, and invests it into R&D. Of this amount, roughly 80 percent is then put into software. However, the development of software in Huawei's early period had relied heavily on one or two geniuses on the team. This meant that things were changed as necessary as the process went along, and if the whole thing failed, the team had to start again at the beginning. The R&D procedures focused on results, not on a structured underlying process. This not only made Huawei's R&D lack a sense of direction but also often led to enormous waste. It made it impossible to safeguard the investment.

In global terms, the United States stands first in software development, and India comes next. The personnel cost of software development in the United States is prohibitive, however. For Huawei, it was more cost-effective to choose neighboring India for its efforts. In addition, the process of having software developed in India would enable Huawei to gain technology more easily. Starting in 1999, therefore, Huawei invested USD 400 million in an R&D facility in Bangalore. It hired more than 5,000 software professionals who worked together with Huawei's team, shoulder to shoulder, in what was called the mighty CMM movement. After three years of hard work, the genes of CMM were transplanted into the bodies of Huawei people.

In June 2001, after getting official international CMM Grade 5 accreditation, the Huawei-India Research Institute launched a process of setting up similar institutes in 17 other countries and cities in China, including Beijing, Shenzhen, Shanghai, Nanjing, Xian, Chengdu, Wuhan, the United States, Sweden, and Russia. After years of effort, Huawei had finally become a multinational company with CMM Grade 5 international accreditation with globalized R&D.

Having adopted these two management changes, which thoroughly reshaped the company, Huawei now gradually found that it had an R&D model that was cutting-edge on a global basis. Its R&D results began to approach European and American telecom standards, while it still enjoyed China's innate advantage in low-cost labor.

In 2004, Huawei won three projects to set up backbone transmission networks throughout all of France and also major cities in Holland and Germany. After this happened, a well-known telecom company in Europe set up a team whose main task was to collect, investigate, and analyze information on the cost differential between European telecom companies and Huawei. After a year, the resulting report revealed data that were shocking enough to stir up considerable alarm in Europe.

In terms of costs per person, Huawei's R&D staff earned an average of USD 30,000, while in Europe, the average income of a telecom R&D person was USD 150,000. In terms of work time, the average number of hours a person in European telecom companies worked was around 1,700, (over a year), after subtracting required holidays, paid leave time, and other benefits. In Huawei, the average number of hours exceeded 3,000. The company implemented an annual leave system, people in the R&D department worked a six-day workweek, and on average each person worked three hours of overtime per day. There was therefore a five-fold difference in salaries and a two-fold difference in amount of time worked. The investment in R&D came to a full 10-fold difference. That is, for every one dollar of investment that Huawei put into R&D, European telecom companies had to invest ten dollars.

These things constitute the so-called secret behind Huawei's ability "to use the small guy to beat out the big guy" and "to use the weak to beat out the strong." Once the board of directors of this European telecom company saw the results of this report, it called an emergency internal meeting and made the following declaration to staff: "Huawei has already become the largest threat to our existence. If we do not find a new growth point, we will have nowhere to go but down once that company truly emerges."

CHAPTER FOUR

THE POLITICAL INSTINCTS
OF AN ENTREPRENEUR

No matter how capable a person, he still has to win the approval of the society within which he lives if he is to have even a chance at success.

—REN ZHENGFEI

In a meeting of the marketing department of Huawei, held in December 1995, Ren Zhengfei made a significant statement: "Huawei went into the telecom field out of pure ignorance. Only after we had taken this road did we find out how narrow the market is, how stringent the demands on technology, and how ferocious the competition."

"When we were starting out, telecom products were highly technical and the profits were high. This made all the giants in the field concentrate in this arena. In fact, it was a competition among oligarchs. We were just a small ant standing at the foot of elephants. We said we wanted to be big too—but this was easier said than done."

Clearly, as a *minying* company, Huawei would have been obliterated as it stood amid a forest of oligarchs, among fierce competition and in a turbulent and dangerous environment, without some kind of strong backup force. In the early period of founding the company, during an extremely difficult time, how did Ren seek government support in order to improve the environment for developing the company?

Being in Step with Mainstream Society

Fees for installing telephones in the 1980s were extremely high, as everyone in China knows. This was due to the high cost of telecom equipment, which itself was the result of China's telecom market being monopolized by foreign companies for years. The percentage of people in China who had telephone lines was extremely small, which meant that 90 percent of telecom operators in China were operating at a loss.

Shenzhen then became the "open window" for China to the outside world. With its preferential policies, it became a kind of experimental plot for the country, and it swiftly rose to become a modern city. The massive wealth that accompanied this process enabled numerous entrepreneurs to be seduced by a life of luxury and dissipation. Although Huawei was small at the time, with revenues that were not in any way comparable to those of the larger Shenzhen companies, Huawei did have core technology, it paid taxes, and it had fairly respectable profits. It thereby attracted the attention of Shenzhen's municipal government. This government recognized that Huawei was a new breed of animal. It had self-generated R&D, a system of shareholding by employees, and what was described as a "wolflike" culture. The Shenzhen municipal government could see that Huawei would become an extremely effective high-tech company in the future.

On August 18, 1996, an official in the Shenzhen municipal government made the following comment:

In the early days, Shenzhen's private sector displayed two main tendencies. One was to squander money. There was plenty of it, so people spent it on buying real estate, high class cars, wining and dining, and so on. The other was to shift the loyalties of a company abroad in an effort to find bigger backing from outsiders. Huawei was different. It quietly worked away and put its money into R&D as well as paying taxes to the State. It did not use its money on private consumption. Instead, it actively tried to figure out how the company could achieve sustainable growth through effective driving mechanisms. It tried to figure out how to participate in international competition. We recognized this about Huawei.

Once Huawei achieved breakthroughs in program-control switching technology, and then went on to make breakthroughs in digital telecom, broadband, GSM, and a series of other core technologies, it was simply logical that the government should extend some support to the company. The company represented precisely the right kind of "new force coming to the fore." Shenzhen's government provided assistance in the form of moral support and macro policy support.

For example, on June 1, 1996, the government decided to provide Huawei with government-backed credit guarantees. Huawei was in a transitional period, in the midst of upgrading its R&D levels and expanding production as it tried to break into overseas markets. Since it was a *minying* company, not government-operated, however, it was being discriminated against as it tried to get loans. In backing it, Shenzhen officials noted, "So long as Huawei can break into international markets with its switches, we certainly will provide buyer-side credit." Not long after, all of Shenzhen's large banks began to make loans to the company. The bottleneck in funding, which had plagued Huawei for years, disappeared. As though it had shucked off a heavy burden, Huawei now entered an era of high-speed growth.

As another example, in December 2009, the government supported Huawei by providing it with a patch of ground that was five square kilometers. This then came to be called *Huawei City.* This decision came after the Shenzhen municipal government carried out a study of Huawei's base at Bantian. It felt that its surrounding environment and the urban facilities were not appropriate to the headquarters of a multinational. Because of this, it decided to adopt what it called the *Toyota model.* Back in 1959, the Japanese city of Jumu provided Toyota Automobile with a manufacturing base, which it called *Toyota City.*

In similar fashion, the municipal government of Shenzhen also provided other *minying* enterprises with support and assistance, including the telecom manufacturing companies Julong, Datang, and Zhongxing. By now, however, some of these other companies have either shut down or been unable to compete with Huawei owing to poor management and lack of vitality.

Clearly, a major consideration for any entrepreneur who is seeking to grow his company is how to get support from the surrounding environment. This means taking advantage of policies in order to grow, but also not relying on policies in order to survive. It involves ongoing technological innovation and a spirit of contributing to the nation so as to be in step with mainstream society. It means being proactive in optimizing the company's internal and external environment. All this is fundamental to safeguarding a company's vitality and ongoing growth.

In this respect, Ren displayed the long-range vision of what could be called a commercial politician. He had an acute sense of positioning. Ultimately, the support and assistance that the government provided to Huawei were not, as rumor has it, due to political motivations. It was due purely to commercial behavior. The key to it was Ren's determination to forge ahead, his spirit of getting things done, and his sense of ubiquitous crises. The examples below will explain this last point.

The Source of Ren's Unwillingness to Grant Interviews to the Media

As Huawei became more well known in China's high-tech fields, the media began to pay ever more attention to its gatekeeper, Ren.

On July 6, 1996, a director from China's CCTV television came to Huawei to shoot a documentary on how science education would help revive China. The documentary team was taken around by Huawei's vice presidents, Zheng Baoyong, Li Yinan, and others, but Ren also granted an interview to CCTV. This documentary caused a sensation once it was broadcast on the *Half-hour of Business* program. After this interview, however, Ren simply disappeared from public sight. This was wholly unexpected. For a long time, not only have journalists from the most influential domestic and international media been unable to interview him in person, but also nobody has been able to catch even a glimpse of his activities. Without exception, people have come back empty-handed.

In 2000, *Forbes* magazine decided to publish a ranking of China's richest people in that year, in a global release. The Englishman in charge of the ranking, whose Chinese name is Hu Run, made the assumption that Ren held at least 5 percent of the shares of Huawei. Using the PE ratio of Cisco and the operating revenues of Huawei in that year, he came up with a valuation of Ren's supposed shareholding. He figured that Ren should be worth around USD 500 million. Based on this, he placed Ren just behind Rong Zhijian and Liu Yonghao, as third in the ranking of China's richest people. Hu Run then sent this to Huawei for confirmation of the data. He unexpectedly received a terse and strongly worded letter from a lawyer in response. What's more, several days later, a lawyer showed up at his door, demanding in no uncertain terms that Ren be taken off the *Forbes* list.

In April, 2005, *Time* magazine voted on the world's 100 most influential people. Ren was the sole entrepreneur from China in this list, put in the same ranks as such commercial giants as Bill Gates and Steve Jobs. Journalists interviewed the great majority of people on the list, and American journalists came to Shenzhen as well but were unsuccessful in gaining an interview with Ren. Not long after, an article appeared in Huawei's internal website about this, called "A few comments on Boss Ren Zhengfei's being included in *Time* magazine list." It said, "What should you say when interviewed by anyone in the media? Say that you are afraid this is overstated and doesn't match reality. Say that quite possibly people don't believe it anyway. In fact, they might consider it false. But best not to see media at all."

A number of senior managers in the company asked Ren to reconsider his approach. They got the reply, "Do you all want to eat well, or are you just interested in fame?"

"Huawei is not a listed company," Ren continued. "It has no obligation to reveal corporate operations or conditions to the outside world. Anyone who violates discipline in this regard will be punished!"

With that, Huawei firmly closed its doors to the media and anyone from the outside world. It has maintained its distance for a long time, not even cracked the door the slightest bit to let anything out. As the company internationalized, it necessarily had to move from being a closed system to being more open. Nevertheless, Ren has, as before, maintained his posture of being cut off from the world, until his gradual appearance in 2015.

People who understand Ren are quite clear about his way of thinking on this. Huawei's internationalization was launched in the midst of very tough conditions. The company's very bones were extremely vulnerable and its survival precarious in an extremely hostile environment. It also had no successful model in front of it to serve as an example. Its competitors, meanwhile, were globally famous telecom giants. Under Ren's command, Huawei not only had to struggle hard over a long time, but also had to grope its way forward in terms of both core technology and management systems. It constantly had to reinvent itself and innovate if it was to avoid being washed out. In all this, it had to accommodate itself to the requirements of changing international competition.

Of critical importance is that Ren's ambition has been to forge Huawei into becoming a world-class company. He has saved an enormous amount of time by keeping himself aloof from the outside world and by refusing contact with the media. He has been able to throw all his energy into reforming management internally and becoming international externally. As a result, it has taken Huawei only 20 years to emerge as a player on the world's high-tech stage.

"It is better to work quietly than to talk too much." This is an iron law of conduct to Ren and one he has assiduously managed to keep. In this, as a world-class entrepreneur, he has been amazingly similar to IBM's previous president, Louis Gerstner, who was president from 1993 to 2002.

CHAPTER FIVE

LEVERAGE: FOUR MAIN FULCRUMS FOR INSTILLING CORPORATE CULTURE

Resources can always dry up. Only culture lives on.

—REN ZHENGFEI

The corporate culture of a company will always reflect the world view and set of values held by its founder.

With Huawei, the thing that troubled its competitors most was its martial spirit, the high degree to which its "wolf forces" marched in step. "In victory, we raise a toast to one another, in defeat, we fight to the death to save one another." "When the going gets tough, it is the brave who win the day." These thoughts had long since seeped into the soul and blood of Huawei people. They are what allowed Huawei, when it was just starting out, to dare to challenge competitors several times its size and to use third-rate products to take over first-rate markets.

The other thing that struck fear into the hearts of competitors was the way Huawei people were willing to "use 10 years to forge a single sword." Core technology was the single largest issue keeping Chinese enterprises from emerging as a group. To confront this issue, Huawei people kept sleeping cots under their desks. When they were tired, they would pull these out and take a nap, then get up and continue working. This kind of "cot culture" enabled the company to complete in 20 years what it had taken the advanced countries in Europe and America some tens of years, even 100 years, to do, namely, accumulate a body of technology.

Patents can serve as an example. In 2000, Huawei applied for just one patent. By 2008, the company was applying for patents daily, at a rate of five to seven inventions per day. It had amassed 35,773 international patents by then; not only did it hold one-half of the international patents applied for by China at the time, but also it was beating out the "old guard" in patent applications, namely, Bell Labs, Siemens, Matsushita, Phillips, and Toyota. It had become the "new aristocracy" when it came to global patents.

This corporate culture, created and guided by Ren Zhengfei, allowed Huawei to overcome numerous problems. In addition, given the extreme scarcity of material resources, it allowed the company to create what were regarded by the Chinese people as astonishing feats. It enabled the Chinese people to recognize that there can be a force that is not afraid and not going to be defeated: "We should not fear the gap between us and others; what we should fear is not having the will to fight." That legacy, brought down to the Chinese people through the ages, is what drives Huawei. To this day, the corporate culture of Huawei deeply influences its 170,000 employees and has become their spiritual totem.

Well, then, exactly where does this martial spirit come from? Where did this force to execute, to get things done, begin? And how did the company actually put into effect this kind of corporate culture?

Those Who Live in Trepidation Survive

The path that successful entrepreneurs take in going from China to the world stage is different for each person. All have one thing in common, however—they maintain a healthy regard for potential disaster. The stronger their sense of potential disaster, moreover, the longer their corporations have a chance of surviving. This law of survival has long since been generally recognized around the world.

Bill Gates is known to have said, for example, that "Microsoft was only 18 months away from bankruptcy." Louis Gerstner of IBM declared that "long-term success is only possible if we constantly keep in mind a sense of alarm." Intel's chairman of the board, Andrew Grove, has even summed up his experience of running Intel for 30 years in the statement "Only those who live in trepidation survive."

"Crises, reductions, and even bankruptcy will inevitably come to Huawei." "Crises are not what mature a company, they are what kill a company."

Ren 's sense of potential disaster is no different from that of all first-class entrepreneurs around the world. What has been different, however, is the material he had to work with. Unlike in developed countries, China's recent history and its massive population meant that resources for education were extremely limited. This led to a caliber of people and level of professionalism that was relatively low. This was particularly evident as Huawei attempted to internationalize. In trying to lead his troops out of darkness and grope a way toward the light, Ren understood quite clearly that his own highly refined sense of potential disaster would be insufficient. He had to take advantage of other measures, create crises artificially in order to let that critical understanding penetrate every single employee's nerve endings. In so doing, he would strengthen the immunity of Huawei and reduce or push back the arrival of real crises.

Crisis management, therefore, was not just the heart of Ren's management thinking. It also became a highly valuable business concept that he brought to all of China's business community.

As everyone knows, Ren is the initiator and promoter of mass movements within Huawei. The purpose in mobilizing people to undertake "movements," however, is not to "rectify" people. Instead, under a specific historic environment, it is to create artificial crises and, in so doing, to upgrade and push forward advances in management.

"Only by designating a battlefield can you form a management team with the same set of values and ability to lead. Only then can you push forward progress in the company on a large scale, instead of having leaders cancel each other out." This statement by Ren describes the core reason he mobilizes people in mass movements.

In the 1990s, China's telecom market came up against a historic opportunity, namely, the country's investment of a very large amount of money, billions of RMB, in telecom infrastructure. After Huawei's revenues exceeded RMB 1.5 billion, its level of management and operating concepts faced new challenges. To deal with those, on December 26, 1995, Ren mobilized a mass movement called a *collective resignation* in the marketing department. All the formally employed cadres in the marketing department submitted their resignations, en masse, together with reports detailing their own considerations. After one month of internal corporate hiring for positions, and a process of performance evaluations, 30 percent of the old guard and those who had "rendered meritorious service" were indeed forced to leave their former positions. During this movement, a slogan was put forward to the effect that "the heroes are those who can sum up the lessons of failure, and then move toward victory." Many cadres then wrote articles about how they felt about Huawei's methods. They put forth ideas about how things should operate and how certain things should be rejected. Overall, this allowed the company to move into a new period of historic transformation with a whole new aspect.

Behind this major move of a collective resignation, described by Ren as "something that shocked heaven and earth and made the spirits weep," was a farsighted strategy. He wanted to make cadres know that they could be promoted but also demoted. Wages could go up, but they could also go down. In letting them know this, he gave them a sense of crisis.

After this, Huawei's high-speed growth brought with it new and ever more complex problems. The mass movements at the company also continued to get more acute and more turbulent, as fuel was added to the fire. A "major discussion of the *Huawei Basic Law*" began, for example, which by now had been in place for three years. In 1998, a movement called "opposing naiveté in developing products" was launched; in 2000, a movement called "internal start-ups" was launched; in 2007, there was another collective-resignation movement, this time undertaken by close to 10,000 old-time employees. This shocked not only the business community within China but also the media. Each mass movement had different methods and components. The thing they had in common was that Ren manufactured artificial crises in order to root out chronic problems that obstructed further growth. His aim was to unearth problems at deeper levels, including declaring war on corruption, completing a handover from old to new, and so on.

In 2000, in a ceremony commemorating the fourth anniversary of the collective resignation, he gave a speech entitled, "Summing Up the

Lessons of Both Failures and Victories as We Recreate Ourselves." In this, he presented his evaluation of Huawei's mass movements.

"Any nation on earth, and any organization, will see its life come to an end if it does not metabolize the old and replace it with the new. If we cling to the past, we will send this company to its very grave. If we had not had the collective resignation of the marketing department, and the impact that had on the company, we would not have enabled any new management advances or new systems to take root at Huawei."

This therefore can be regarded as one major "fulcrum" by which Huawei's corporate culture is actually put in place.

By now, we can look back and see that Huawei's mass movements had a major role in unifying the thinking of employees over the past 27 years. It also had a major impact on strengthening the cohesiveness of the enterprise and its will to fight. Even more critical, however, Ren's mobilization of mass movements instilled an awareness of crisis into several tens of thousands of young employees at a time when Huawei was facing the arduous task, under unique historic circumstances, of going from being a small Chinese workshop to becoming a world-class corporation. This device enabled Huawei to grow in the midst of tough times. Indeed, it made Huawei "big and strong" in the midst of tough times.

Self-Criticism

If mass movements enabled Huawei people to feel a sense of crisis, self-criticism enabled them to recognize the gap between themselves and others. Once any person, or any organization, has an accurate recognition of its own smallness, behavior can then begin to move in the direction of strength.

In September 2000, more than 3,000 R&D personnel participated in a large conference. The company had pulled together a massive amount of materials that were sitting in inventory and that were unsalable because of "blind" R&D or mistakes. It also pulled together quantities of plane tickets and receipts for hotels, invoices for compensation for contract violations, and so on. It put these in specially made frames which it then called "product awards." It issued these for several hundred relevant

R&D personnel to take a look at. At the conference, Ren delivered a report that he called "Why We Need to Criticize Ourselves."

"Huawei is still a very young company," he said. "Even though it is full of vitality and enthusiasm, it is also absolutely full of naiveté and arrogance. Our management is not in any way standardized. We will grow up more quickly only if we recognize this and constantly critique it. We do not criticize purely for the sake of criticizing. We also do not criticize in order to repudiate everything. Instead, the aim is to optimize and to build up, the overall goal is to lead us in the direction of upgrading the entire company's comprehensive core competitiveness." He continues:

> *We are in an age when the IT industry is changing dramatically, an age that moves at 10 times the previous pace. The only thing that doesn't change in this world is change itself. If we are just a little too hesitant, we fall behind by a thousand li. If we are complacent and conservative, reject criticism, then we fall behind by more than a thousand li. Should we move toward failure and death just for "face"? Or should we forget about face, cast off mistakes, and turn ourselves in the right direction? If we intend to survive, we must "surpass," and that means going above and beyond. The necessary condition for "surpassing" is ridding ourselves of mistakes in time. To get rid of all mistakes, we must be courageous enough to criticize ourselves.*

From time immemorial, both inside and outside China, self-criticism has been regarded as one of the sharpest "thought weapons" by which humanity can propel itself forward. However, in China, to have employees use this very efficient weapon to carry out "self-repudiation" and "self-dissection" is absolutely no easy matter. One can only imagine how difficult it was in a large-scale, high-tech group of people with 170,000 intellectuals, all with higher degrees who were particularly concerned about face. Inside the company, Ren pointed out numerous times that "People are afraid of pain. And the greatest pain is that which you inflict upon yourself." Therefore, he advocated self-criticism but he did not advocate mutual criticism. He felt that it is hard to get measurements right in criticism—too little, and it is not effective; but too much, and the medicine becomes too "hot." It then is easy to create contradictions within a team. Self-criticism, however, ensures that everyone holds back on the severity, goes a little easy on himself. Even if you use just a pencil to beat yourself lightly every day, that's better than not beating yourself at all. After a few years, people can be forged into steel.

This can be regarded as another major "fulcrum" by which corporate culture at Huawei was actually put in place. Ren's task then was to turn this fulcrum "point" into a whole "sheet," which really achieved wholesale change. In taking the lead in this, he also established an example for the rest of China.

For example, in a high-level meeting of cadres "democratic life," he undertook to self-dissect himself as follows: "I don't understand reporting forms, and I don't understand management. My only advantage is that I can change when I am wrong, since I don't pay much heed to this concept of face." "Face is a shield by which incompetent people protect themselves. What outstanding men and women should pursue is truth, not face."

In 2000, problems cropped up with respect to xiaolingtong[1] and code division multiple access (CDMA). Ren took responsibility. "In Huawei, I am the one who has made the most mistakes of all." He also delivered a document to himself, lowering his salary. In 2008, once he became aware that the innovation movement he had launched within the company was in fact creating a great deal of insider transactions and corruption, he took the lead in pulling his investment out of the companies involved, if they were owned by friends and family members. Moreover, he proposed that the Executive Management Team (EMT) institute a collective oath-swearing system that, in terms of "anti-corruption and pro-clean" aspects, would be subject to regulatory oversight by the entire body of employees.

In addition to teaching by personal example, when it came to promoting cadres and evaluating their performance, Ren said quite explicitly on several occasions, "The company believes that self-criticism is a good

---

[1]Xiaolingtong was a Chinese technology that was halfway between mobile and fixed, and far cheaper. Uncertainties about what technology China would use, and who would get licenses for the technology, meant that suppliers had to hedge their bets.

way to further an individual's progress. It hopes that all departments, at all levels, will not promote those employees who have not grabbed hold of this weapon." In 1998, when the draft of the *Huawei Basic Law* was being prepared, a large conference was held for officials above the department level in China Telecom and China Unicom. Ren gave a speech at this event, to which he gave the title "Huawei's Core Values." In this speech, he even declared that the basis for Huawei's long-term stability is "whether or not successors have this spirit of self-criticism."

Ren took self-criticism to an extremely high level, and it did indeed play a pronounced role in furthering Huawei's progress and development. However, self-criticism also implied rejection and repudiation, whereas innovation and creativity required risk taking. Many cadres therefore became afraid of making mistakes. They felt the wiser course of action was self-protection. Ren drew blood on this issue when he said scathingly, "Everyone who aims to protect his own self-interests will have his job eliminated. He is already a stumbling block to change."

He added, "Is it possible to take away somebody's position if he hasn't made any mistakes, and also hasn't made any progress? Well now, if somebody says he has not made mistakes, how can he be a cadre? Some people have indeed never made a mistake, and the reason is that they have never done a single thing."

When Ren evaluated people for promotion into the ranks of core cadres, he first took into consideration those who had indeed made mistakes but who had corrected them and moved on from them. He felt that setbacks and failures were a precious form of spiritual wealth. The lessons of this kind of experience not only were beneficial to the person himself, but also could warn other cadres not to make the same mistake. It could help them make fewer wrong turns. Clearly, Ren had good intentions in wanting this critical understanding of self-criticism to work its way into people's minds. He did not encourage cadres not to make mistakes. Instead, he wanted them not to be afraid of mistakes, while at the same time being brave enough to correct and revise mistakes and draw nourishment from the process.

To institutionalize this launch of self-criticism, one important move Ren made was to set up a "self-criticism guidance commission." Its most important function was to guide and help each department in setting up an organizational atmosphere of self-criticism. At the same time, Ren mandated that the journal called *Huawei People* have a regular column on self-reflection. When senior cadres discussed their own faults and mistakes at "democratic life meetings" and carried out self-criticism, these would be published openly in a series of columns. Meanwhile, a number of articles exposing Huawei's management defects also began to be published as a series in this journal. In a top-down effect, what had been mobilized by senior-level teams was then taken up at middle and lower levels of management and among regular employees. A general mood then formed at Huawei in which all cast off "face." Through the method of democratic life meetings, all busily began to self-reflect on their own work, digging out inadequacies and correcting defects in a timely manner. After sustained promotion of this process over many years, an institutionalized spirit of self-criticism did indeed permeate people's attitude.

"Self-criticism is not self-castigation but rather self-confidence. Only the strong can do it. And only if you in fact do criticize yourself will you be strong. Self-criticism is a kind of weapon as well as a kind of spirit."

"Only those who adhere to self-criticism over a long time have a broad-minded approach. Only companies that are critical of themselves over a long time will have a bright future. Self-criticism has allowed us to arrive at where we are today. How far we can go in the future depends on how long we can keep self-criticizing."

Ren made these comments on September 2, 2008, at an awards meeting for core technology in network products. They can be seen as a crystallization of his experience in governing Huawei. In fact, if one were to say that Ren had any magic potion in putting Huawei on the international stage, this would be it: self-criticism. Due to self-criticism, Huawei people saw clearly where they had problems and that they needed to adopt measures to correct them. As their ability to correct mistakes grew stronger, the muscle of the organization became healthier. The healthier the organization became, the better it was at doing self-criticism.

By practicing this virtuous cycle and continually replacing the old with the new, Huawei evolved along the difficult path of becoming a multinational.

We Will Take Care of Those Who Contribute

China does not lack for those who make contributions to the nation, who pay "tribute," so to speak. What it lacks is the platform on which, and the mechanisms by which, these people can really rise to their potential. In Huawei, success and business results signify wealth. As long as you make a contribution, you need not worry that you won't be compensated for it. What's more, you will be rewarded handsomely. "Only when people see that those who contribute are truly rewarded will others stand up and fight to be productive as well." Ren's confirmation of this principle has served as a powerful driving force in motivating people.

In fact, "We will take care of those who contribute" appears obvious and easy to understand, but it is extremely hard to implement. To put it into effect, in 1998, at a human resources meeting, Ren made an important determination in this regard.

Performance appraisals are made by human beings. Even though the members of evaluation committees are as fair as possible, they still are humans. They are living, breathing, blood and flesh and it is hard for them to put this fact aside when they make evaluations. Their views will always be circumscribed by certain factors. Moreover, they can never make absolutely everyone satisfied with their appraisals.

If the company is to develop swiftly, we cannot wait for all results to come in before making final judgments. Each stage of an evaluation will necessarily have aspects that are incorrect. What we ask is for each rank or grade in the company to be as fair and just as possible. If we cannot do that, this company will have no future.

This is the principle by which Huawei evaluates performance, a fair and just approach that "seeks truth from the facts." It also can be regarded as the compass that guides thinking on the question of how to actually implement Huawei's promise "we will take care of those who contribute."

For example, in terms of fairness, Huawei has a hard-and-fast rule. One week after coming into the company, every single person is regarded as being on the same starting line. That goes for whatever the person's academic record, PhD or Master's or Bachelor's, and it goes for whatever his honorary position—none of this counts any more. All start out as "workers." All obtain opportunities to grow based on their own actual experience in the company. Those who are willing to put their heads down and make a contribution are those who create their own opportunities, their own stage for advancement. This means that the entire organization is forging ahead in a way that understands "one can go up, but also down." This rule operates in a way that requires fair competition.

In terms of justice, every department in Huawei has a hard-and-fast rule about washing out the bottom 5 percent. Since the competition is extremely brutal, this system reflects Huawei's principle of fairness and justice. It is hard to avoid having a few "clever people" who loaf around while tooting their own horn and who thereby impact the results of the entire team, when teams are made up of highly intelligent and highly competitive commercial instincts, people who are also highly compensated. In combating this tendency, Ren uses a West Point practice to warn Huawei people: "We wash out those who are last-enders, at the end of the line. We cut out those who don't work hard, or employees who simply cannot handle the work. The reason is that this is a way of stimulating the organization. At the same time, it is a way of protecting the more outstanding employees."

To implement this philosophy in a very practical way, to keep those who do make a real contribution from being attacked and excluded, and to avoid the malicious slander of those who can't actually do much but who are very good at manipulating others, Ren personally proposed a system of employee forums for employees "who are put in an adverse position." By this he meant people who felt they were unjustly singled out. Anyone who is in danger of being washed out and who does not accept the results of his performance appraisal can appeal his case. To this end, Ren set up an honorary department composed of "elder experts" who served in political work departments for many years. Their primary function is to resolve disputes relating to performance evaluations. Once verdicts are determined, they are to help employees who have been fired for just cause, submit to this both mentally and verbally.

The honorary department does not accept claims that are put forward by departments. It only accepts suits that are initiated by individuals, but the claimant must use his true name. The idea is to keep people from employing fraud and deception in order to damage others. Once a case starts in which an employee does not submit to the results of his performance evaluation, the elder experts take on the case and begin to do a detailed investigation to understand the entire process. They look into the capabilities of the employee, his hidden qualities, and contributions he has made in the past. Through a comprehensive approach, in the end they make a determination about the performance evaluation results themselves.

Because of this group of elder experts, the results of performance evaluations at Huawei are never the last word. They are never the "single blow that hits the final note." Many cadres in Huawei are quite young but hold tremendous power. Nevertheless, they generally are not arrogant and willful in performance evaluations. In this way, although not every single performance evaluation is totally fair and just, in general they are indeed objective and "seek truth from the facts." When minor mistakes are made, they are adjusted through the process of internal democratic life meetings and through communicating corrective action in gentle ways.

Huawei implements another system that is called a counselor or guidance system. It has set up this system to protect new employees and ensure they don't take wrong turns. The aim is to make sure they acclimate to the company as fast as possible and understand its organizational atmosphere and principles of fairness and justice, with the ultimate goal of ensuring that they can be as creative and dynamic as possible. Once older employees have been in the company a given number of years, they are required to take on new employees as their wards for three to six months, in order to guide them, help them, and lead them. If a new employee is washed out within three months, then the guidance counselor must bear some of the blame.

The guidance counselor system forces counselors to do all in their power to guide new employees properly. This ensures that activities in which one helps another and both "both become excellent in their work".

Starting in 1996, Huawei spent a large sum of money to implement the Hay compensation system in a systematic way. Adopted from the United States, this system also is used internationally to determine job qualifications and compensation packages. Huawei used it for a number of years, refining it for its own purposes. In 2002, on the basis of this system, Huawei designed and began to implement a job qualifications system and a "five-grade two-way system," that is, performance evaluation mechanisms that combined both eastern and western practices. As a result, Huawei has institutionalized the more nebulous aspects of performance evaluation principles that are meant to operate according to fairness and justice and seeking truth from the facts.

Huawei has gone through immense cultural transformations, from a wolflike culture to an "everybody rowing in unison culture," to a culture that is "customer-centric" and one that "gives primary consideration to those who fight." None of these changes has modified the underlying guiding principle of the company, namely, that it will take care of those who contribute. This in turn is what allows the corporate culture to become so embedded in people's minds.

Clarifying the Question of Who We Are Fighting For

The incentives provided by the principle of taking care of those who contribute allowed outstanding employees to come to the fore. As Huawei moved forward briskly, its employees made money. At the same time, however, Ren discovered that while money could generate powerful consensus, loyalty, and ability to run the operation, it could not generate trust and belief. In the process of going from a six-person workshop to a multinational with 170,000 employees, all those entering the company had different values, different mental horizons, and different life aspirations. Some employees were working purely for money, while others were more idealistic. The dilemma Ren now faced was how to incorporate the company's own core values into the beliefs of all employees. As Bill Gates has also said, the issue became how to create enduring or sustainable incentives.

American scholars have researched this subject in depth. The two authors of the book (Jim Collins and Jerry I. Porras, *Built to Last: Successful Habits of Visionary Companies*) spent three years studying 18 corporations that have all been going for a long time, some close to

100 years. They include GE, Walmart, Disney, and others. After comparing these companies to the rise and fall of their competitors, the authors were surprised to discover that the one thing the 18 corporations had in common was a "religious-type superlative corporate culture." This had a massive influence on stimulating the growth of professionalism in these enterprises.

China stands in stark contrast to this. Historically, the country has been mainly agrarian. It relied on "the heavens" for sustenance, and the clan was the core entity, so what emerged was a Confucian culture that regarded blood as being thicker than water. This culture deeply penetrated people's minds. To a degree, the individual effort that was fostered by this kind of small-scale agriculture prevented more collective efforts and the development of advanced professionalism in Chinese enterprises. Since Ren set his sights on making Huawei a multinational company, naturally he was extremely attentive to this whole issue. In describing the importance of Chinese culture to the development of Chinese corporations, he had the following to say:

"The Chinese culture has endured for some 5,000 years, which is an indication of its resilience and its durability. However, the fact that the motherland has gone through such a long history and yet not 'arisen' is not because Chinese people are not worth much. Instead, it is because China never generated large-scale industry. To this day, the individual efforts fostered by a small agricultural mentality have not emerged from the labyrinth, the maze. This has even led to that saying that Chinese people are mere worms."

Objectively speaking, blood relationships and the ethical system of Confucianism have run through the culture of China for 5,000 years, and filial piety toward parents and concern for brothers and sisters are what have been able to consolidate that whole system and create a solid core. In Huawei, to guide employees out of their nest of individual efforts and strengthen their awareness of collective efforts, the idea now was that you had to start with small things as you nurture and train employees. First, they must learn how to love themselves and their families. Then you extend that kind of love toward the company and colleagues. Finally, you raise it to the level of loving the motherland, loving the people. This is the ultimate realm as expressed by the Confucian adage, "Cultivate one's own self, unify one's family, govern the country, make all under heaven at peace." It was also the mental compass by which Ren wanted employees to "clarify to themselves whom they are fighting for."

In the past, China's cultural orientation was something that was instilled into every child, starting from primary school: "Everyone must be concerned about the great affairs of the country." As a result, everyone was indeed concerned about the great affairs of the country, but nobody worried about how to do her or his own job very well. Huawei was precisely the opposite. It did not encourage employees to worry about the nation's major concerns. Instead, it encouraged them to do their own jobs well. Inside the company, Ren repeatedly taught employees this lesson as he said, "When we say that everyone should 'seek truth from the facts,' we mean they should do the little things well. Once those are consolidated, they become big things. If you do your own job well, the company will prosper and its contribution to the nation will be bigger. It will then be easier for the country to handle bigger affairs. In this way, individuals realize their own ideals, and the company fulfils its responsibility to society."

"Use material civilization to consolidate spiritual civilization. This refers to what we call the two engines. One is the country, the other is oneself. In subjective terms, you are working for yourself; in objective terms, your work is for the country."

By looking back on Huawei's 27 years and the fierce conflict between eastern and western cultures as the global economy became one unit, it is clear that collective struggle is a quality that Huawei cannot do without as it emerges onto the world's high-tech stage. By clarifying whom we are fighting for, Huawei has been able to generate an unimaginably powerful force of mental energy. This is also the key to how such statements as "In victory, we raise a toast to one another, in defeat, we fight to the death to save one another" and "Share a bitter hatred of the enemy" have truly been planted in people's minds and have taken root. The paths of Huawei and western corporations have been different, given that the Huawei spirit differs from the spirit of professionalism that evolved out of western religious denominations, but both have similar internal driving forces.

PART TWO

CHANGE

If one wants to understand how Huawei evolved from being a small workshop to a multinational, one must understand the forceful measures that Ren Zhengfei has taken, the painful process by which he has "remolded" and "thoroughly washed the heart and modified the mind" of Huawei people, their genes, their organs, their very skeletons.

"Study large corporations as thoroughly as possible. Only then can you be sure to take fewer wrong turns, and pay less tuition." This comment by Ren reveals the core thinking behind Huawei's rapid growth, evolution, and change as it became a world-class enterprise, as it reduced the distance between China and leading enterprises in Europe and America.

CHAPTER SIX

THE *HUAWEI BASIC LAW* AND ITS HISTORIC MISSION

The key to whether or not an enterprise is long lasting depends on whether or not the leadership succession confirms and adopts its value outlook, and whether or not the successor has the capacity to criticize himself.

—REN ZHENGFEI

The year 1995 marked a major turning point, a watershed for Ren in terms of his thinking about management.

In October 1994, Huawei exhibited at the First Beijing International Telecom Exhibition and enjoyed enormous success. Its C&C08 switches began to move into rural markets on a massive scale, and sales volume reached RMB 1.6 billion. Although this seems hardly worth mentioning at a time when China's annual investment in building up telecom came to hundreds of billions, it indicated that Huawei, together with such domestic manufacturers as Julong, Datang, and Zhongxing, had broken through the monopoly that foreign companies previously held over China's market. It also indicated that Huawei was now facing a historic choice in how to become one of the world's top 500 companies.

Which Way Should Huawei Go?

After Ren completed his investigation tour of the United States in the fall of 1992, one of his conclusions was that it was going to be extremely important to improve Huawei's management if there was to be any chance of catching up with the United States. Initially, Huawei's main problem had been how to achieve a breakthrough in core technology. Now that products had begun to be applied on a certain scale in the market, the next most urgent question became how to shorten the management gap between Huawei and the telecom giants in Europe and the United States.

Between 1993 and 1995, Ren led his team on visits to Alcatel, Siemens, Fujitsu, Matsushita, NEC, and Shanghai Bell. In person, he evaluated the world's premier R&D models, management systems, and the apex of the world's production lines. He was astonished time and again, and after returning, he said that phrase about ants and elephants: "We were like a small ant standing at the foot of elephants. We said we wanted to be big too—but this was easier said than done."

In terms of the competitive situation at the time, even though China's market for telecom had just begun, this market for 1.3 billion people was completely occupied by companies such as Lucent, Nortel, Siemens, and Alcatel. Given their advanced technology and massive financial clout, they had already snatched and occupied the world's largest potential market for telecom. What's more, as Julong, Datang, Zhongxing, and other domestic manufacturers were emerging as a group, this led directly to a complete logjam of foreign and Chinese products. Ostensibly, supply could not meet the demand in China for control-process switches, but in fact the situation was that supply exceeded demand.

What's more, Huawei was a *minying* company, not remotely comparable to the telecom giants in Europe and America. Meanwhile, as compared to Julong, Datang, and Zhongxing, with their tremendous depth of state-owned assets and the groups that were listed on the market, Huawei had no advantages to speak of. If Huawei could not occupy a piece of turf within China as fast as possible and "carve a bloody path" for itself in the international market, then with the intensification of competition, the market in China would be saturated and Huawei would be just sitting around, waiting to die.

Perched at this decision point on this continental divide, which way should Huawei turn to reduce the tremendous gap between it and its competitors?

Problems on Top of Problems

When he was founding the company, Ren seized a historic opportunity and swiftly built up the company based on courage and charm. However, because of a lack of experience in management, the organization that gradually came out of the intense process of expansion now faced a new series of problems that were like a tangled skein of hemp around Ren's neck. Faced with formidable competitors, all he could do was strengthen corporate culture and build up spirit in order to increase the corporation's cohesiveness and ability to fight. With these intangibles, he pushed forward Huawei's expansion.

In February 1995, Ren wrote two important documents on how new employees should conduct themselves and on how all employees should behave. One was addressed to new employees, and the other was titled "Rules of Behavior for Huawei People." After this, he wrote another document that was more emotional, called "Spiritual Program." It included this exhortation: "Do your utmost, for your own welfare and your family's welfare, by helping Huawei catch up with and overtake the most advanced levels in the world. Be self-reliant as you pour your forces into setting up and developing Huawei's product systems. Unite with all employees to achieve this goal. Each person must fight hard, go all out to work for the prosperity of the company."

The core of "Spiritual Program" was to bring consensus among all employees about Ren's most ambitious goals—dividing the market into three parts, giving China its due, and thereby unifying thinking, actions, and fighting goals. The question was, Could employees really comprehend the deeper significance of all this?

What's more, during its initial tough period as the company was being founded, Huawei developed a corporate culture that had its own unique qualities. These included such things as taking care of those who contribute, cot culture, wolf culture, and so on. What, in fact, was actually the very marrow of Huawei's bones in terms of corporate culture? What role could it play in driving the company forward in this new moment of history opportunity?

In 1995, Huawei was able to take over a large swath of the rural market, based on its pricing strategy and its strategy of using "oceans" of people.

However, since incentive mechanisms were not yet properly calibrated, this led to chaos at the management level. For example, frontline marketing people in the 32 offices that Huawei set up in all the provinces and major municipalities around the country all worked extremely hard. Each province had different economic conditions and uneven levels of development, however, so each had a different degree of investment in telecom, leading to very different sales revenue in each Huawei office. The question then was how to define results and how to compensate people in a way that would make everyone maintain a fighting spirit and still want to expand the market at a fast clip.

Issues like these forced Huawei's senior management to take into account constantly changing new elements as they considered solutions. Since market conditions were evolving so quickly, however, solutions that actually got passed were soon inadequate to meet the strategic demands of Huawei's explosive growth.

A Glimpse of Light

Professor Peng Jianfeng recalls being invited by Huawei to come to the company to give lectures. He wrote of this in an article called, "The Birth of the Huawei Basic Law." He, Bao Zheng, Wu Chunpo, and Sun Jianmin, were to cover such things as the "second founding" of the company, the shift in its strategy for sales, marketing, and human resource management. This professor from China's Renmin University and the others discussed topics that had a profound influence on Ren, including management aspects relating to this second founding of the company. After the course was over, Ren expressed tremendous appreciation to the professors from Renmin University. He noted that issues to do with China's *minying* companies going through a kind of rebirth were precisely what Huawei itself was facing. He then asked the professors if they could provide consulting services for Huawei, use the company as an experimental plot. He said, "Under your advisory assistance, help us take a small company and turn it into a large corporation. That would be a massive accomplishment both in terms of scholarship and in practice. Between us, we could achieve a win-win situation."

Ren soon called senior cadres together to discuss the second founding issue addressed by the professors. He felt that this was a core issue that Huawei urgently needed to resolve in the midst of its high-speed growth. He hoped that everyone would take it into serious consideration. At the same time, he dispatched the head of the performance evaluations department to meet with Prof. Jianfeng and formally invited the four professors to serve as management consultants. Their advice was to focus initially on human resources and production management and then gradually to extend to organizational structure and building up a corporate culture.

In January 1996, Hong Kong began drafting the *Basic Law of Hong Kong*. In similar fashion, it too began discussions on how the future city could achieve sustainable growth. Inspired by this, Ren glimpsed the outlines of a vague idea that gradually became more apparent. In its "first founding," Huawei had practiced an "extensive mode" of management. This played a certain positive role in developing the rural market. In its second founding, however, Huawei had to beat out Lucent, Siemens, and other telecom giants in Europe and the United States on their own turf. This required management capabilities that were no less than those of their competitors. The urgent matter at hand, therefore, was to have a war plan that could guide Huawei's expansion and then to build on it, one that could gradually enrich Huawei's management systems and operating systems.

Once Ren had settled on this idea, the next step was to decide which person to ask to draft the various systems.

In 1995, a number of well-known international management companies had landed in China, including McKenzie, Boston Consulting Group, Anderson, and Price Waterhouse Coopers. Huawei had some contact with these companies during this period. After hearing a few of their lessons, though, Ren told senior management that he felt their "foreign coaching" about management concepts was overly westernized. Without being guided by Asian culture and grounded in Asian culture, Huawei would find it hard to absorb this coaching. It might even lose its own original culture if it went wholeheartedly western.

What's more, Ren discovered that the professors from Renmin University had considerable overseas experience. They had studied abroad as well as worked abroad for many years. Not only did they have a penetrating understanding of advanced management, but also they had done extensive research into how Japanese enterprises became international entities.

"We hired professors from Renmin University because they understood western culture but because they were also quite clear about Chinese culture. They used western culture to upgrade Chinese culture, which was in sync with Huawei's way of thinking." In a meeting at the president's office, Ren explained his thinking on why he hired these professors to draft *Huawei Basic Law*. In early March 1996, a small drafting group was formally established comprised of Peng Jianfeng as the head, with Bao Zheng and Huang Weiwei as primary authors.

Throughout the next month, Ren relayed to the professors his own personal experience in founding the company, lessons he had learned, and new issues facing the company at its current stage. After extensive exchange of opinions, various ideas came together and the professors sketched out an outline for *Huawei Basic Law*.

Ren read a faxed copy of the outline and felt that it dealt with corporate culture and people's thinking and ideological issues but did not address such things as the relationship between work flow and building the right organizational structure. He therefore made a special trip to Beijing and launched into deeper discussions with the professors on the more broadly based purpose of *Huawei Basic Law*, what its point of view, its positioning, and its method of expressing itself should be.

"The question is how to take our 10 years of hard-earned experience a step further as we absorb the best methodology and thinking in the industry. This 'law' should become theory that can guide us as we move forward, prevent us from falling into empiricism, so that we can find driving mechanisms that enable sustainable growth."

"We want to plant a value system that is up to date in the minds of every person and thereby make sure that thinking is unified between boss and employees, that behavior is unified, so there is a certain psychological contract."

After many such discussions, Ren was able to relay the main thinking that should go into the drafting of *Huawei Basic Law*.

Finally, at his suggestion, six professors from Renmin University abandoned the idea of doing the work in Beijing and began to station themselves at Huawei to do the writing on location. The six professors were Peng Jianfeng, Bao Zheng, Huang Weiwei, Wu Chunpo, Sun Jian-min, and Yang Du.

Order Out of Chaos

Highly adept at both studying and thinking things through, Ren made sure that the office of the people drafting *Huawei Basic Law* was located right next to his. Every time the professors ran into a new problem or came up with new ideas, he could easily talk things over with them. The views that Ren had summarized in the course of his own operational experience helped stimulate the imagination of the professors, while their ideas about management also sparked Ren's thinking about the positive and negative things derived from his eight years of running the business. In China, enterprise management is as yet virgin territory, and so everyone's ideas are fresh and also combative. The friction between the "experience" faction and the "scholarly" faction raised numerous sparks and quickly ignited ideas and expanded everyone's mental horizons.

In an article titled "Order Out of Chaos," Prof. Weiwei recalled the process, how the drafting small group carried out its work. Members of the group did extensive interviews with Huawei's cadres and employees. They asked people to recall Huawei's first founding and give their expectations of the second founding; they combed through quantities of source material, including corporate documents, and combined these with the lessons derived from the tough time of founding the company. Eventually, they came up with key questions: Could Huawei's past experience guarantee its ongoing success? If not, then what new elements were needed? In exploring answers, they combined strategies from China's ancient military works, including the *Sun Zi Art of War*, with the corporate culture, corporate principles, and code of conduct of such internationally famous companies as IBM, Hewlett-Packard, and Intel. They came up with a "core value outlook" that had seven primary components: "aspirations, employees, technology, spirit, interests, culture, and social responsibility." Operating policies were deduced and arrived at, one by one, which addressed Huawei's management of operations, the organization, and internal controls.

On the question of whether human talent was a core competitive strength, for example, after repeated discussions between Ren and the small group, the wording became as follows: "Huawei's greatest wealth lies in its employees, people that the company has been conscientious in training and making effective. Our sustainable growth inherently requires respect for knowledge, respect for individuality, and respect for those who collectively fight for the cause but who are also not excessively accommodating." This then became a hard-and-fast personnel policy for Huawei.

On the question of distribution of profits and benefits, which was a fairly sensitive subject at the time, Dr. Yang Du, who had spent many years studying abroad, did a detailed comparison of Huawei's incentive mechanisms with those of European, American, Japanese, and Korean companies. He looked at Huawei's system of shareholding, its respect for knowledge, and its "collective fighting" mechanisms. After discussions with Ren, Dr. Du courageously put forth the idea of "Knowledge Capital Theory," which was the Chinese takeoff on the word *capitalism*. This was greatly appreciated by Ren. After that, an overall approach to creating value in the company was adopted, one that included labor, knowledge, and entrepreneurship as well as capital. These then became the core components of deciding how to distribute interests. Not only was this system unique to Huawei, but also it became an important part of the incentive mechanisms. It was like a filling station that added fuel to people's willingness to undertake "collective struggle."

An orientation for moving forward was then clarified in *Huawei Basic Law* that was both visionary and highly ambitious. It defined a tight organizational structure and operated through "seeking truth from the facts." It described how to take Huawei through a specific turbulent period in its history, how to move out of chaos into order. Newton defined the laws of gravity, but Watt used scientific keys to invent the steam engine, thereby initiating the industrial revolution in Europe. In guiding corporate management, *Huawei Basic Law* has similar historic significance.

"The 'order' that a twenty-first-century company should have was already coming into existence in China during the dozen or more years in which Chinese corporations have been undergoing reform. However, whether or not China's companies can really move out of a state of utter chaos will depend on the 'initial conditions' that we are able to set up.

It will depend on how our innate character is formed by the directions we choose. In the *Huawei Basic Law*, I fortunately begin to see the answer." This statement was written by the primary author of *Huawei Basic Law*, Huang Weiwei, as it was being created.

Exploring Management at the Highest Realm

Ren has a powerful capacity for self-analysis. This is made evident by his being able to change and move forward on the basis of what he sees around him. He discovers his own shortcomings as he watches the success of other companies, and his own underlying problems in lessons from companies that fail.

One example of this comes from the field of finance. In 1988, a person named Guan Jinsheng founded what became China's largest securities company, called Wanguo Securities. Guan had a double PhD in French literature. He had worked for four years and gradually made Wanguo into a world-class securities company. However, in 1995, in what became known as the "March 27 national bond incident," which he had personally planned, Wanguo lost RMB 960 million in assets in the space of eight minutes.

Ren was highly shaken by this when he read about it in the newspaper, in an article called "How the Number One Securities King Collapsed." He mailed a copy to the drafting group, on which he wrote, "I'm passing this on to you, to let you know why we need to do this *Huawei Basic Law*. Such a likeable man, and so capable, and yet in eight minutes, he sent a world-class securities firm to its grave. Given the pressures, inside and outside, who is to say that something crazy might not happen to me too?! History is a mirror."

The tragedy of Wanguo Securities directly influenced Ren's decision to add a section on control policies to *Huawei Basic Law*, to strengthen the "immunological resistance" of both himself and his management and employees. As the drafting group was discussing the contents of this section, Ren repeatedly emphasized that the key reason many *minying* companies tie themselves up, fail to progress, and even go crashing down is that they do not recognize the importance of constraints. They have too many driving forces and not enough restraining forces. "Huawei's management system must therefore have different levels of both driving forces and constraining mechanisms, in order to ensure that it is orderly, that it moves forward and grows in regulated fashion. We must avoid the mistakes of Wanguo Securities. No matter how prestigious the boss is, there absolutely must be constraints on improper behavior." It was mentioned that while they were discussing the core content of these "Control Policies," Ren asked Huang Wei Wei, "What do you believe is the highest state of management control?" Without thinking twice, he answered, "The highest state of management control is to be able to reach your goal without controls."

"Right! The *Huawei Basic Law* is being put in place in order to reach this condition." He went on in an inspired manner: "Nobody has to tell the waters of the Yangtze which direction to go in, yet the river flows toward the ocean, not upstream. The big bosses of foreign companies are all out there playing golf, while we senior managers here in China exhaust ourselves working all day. The reason is that we have not yet reached that highest state of management control."

Through discussing and drafting *Huawei Basic Law*, Ren was already defining a sustainable path for the company. Exactly how he was going to actually put in place that "highest state of management control" now became the matter he most needed to resolve.

Looking for the "Way" in America

In late 1997, on Christmas eve in the West, Ren again landed on American soil. He brought with him all the questions and concerns about how to make a small company evolve into a world-class enterprise.

People were surprised that on this trip he retraced the route he had taken five years earlier. He also similarly planned to use just one week to cover both east and west coasts of the country. He planned to finish his "investigation" of Hughes Network Systems, Bell Labs, Hewlett-Packard, and IBM. As the R&D center for Lucent, Bell Labs had seven Nobel prizewinners in it. Not only had it invented telecommunication satellites, electronic digital computers, long-distance television transmission, mov-ies with sound, and so on, but also it had proved the validity of the Big Bang theory. It had made outstanding contributions to human progress. As he walked into Bell Labs, Ren revealed with considerable emotion to those around him, "I truly worshiped Bell Labs as a young man, with a kind of respect that transcended love."

In the John Bardeen laboratory, where transistors were invented, which opened the beginning of a new electronic era for humankind, he said with reverence, "John Bardeen belongs not just to Bell Labs but to all of humankind." Ren rarely allows himself to be photographed, but in this instance, he specifically had himself photographed together with Li Yinan in front of the memorial for John Bardeen, as a memento. When this photograph was published in the fourth edition of the *Collected Articles on Huawei*, it amazed and delighted everyone in the company.

However, as the investigation moved into deeper territory, Ren was alarmed to discover that Dell, which had started out at the same time as Huawei, was already a formidable adversary to Hewlett-Packard, which had grown to have an annual output value of USD 3.5 billion. Meanwhile, Cisco had been roughly the same size as Huawei in 1992, yet after five years its output value was over several tens of billions of U.S. dollars. Huawei, which had "grabbed the historic opportunity"[1] and then been through 10 years of steady hard work, had output value of just USD 500 million.

This enormous disparity made Ren become aware of the fact that Huawei was in fact growing slowly. There were both internal and external reasons for this. On the internal side was the lack of knowledge about management. On the external side was the difficulty in getting senior per-sonnel who were of sufficiently high caliber and had international man-agement experience in large, high-tech companies. Attracting one or two such people was not sufficient. What Huawei needed was an entire group of them. In contrast, once an American company had developed a new product, all it had to do was give the word and it quickly could assemble

---

[1]The phrase used here is a classic Chinese saying about licking bitter gall in order to keep yourself awake to study.

a highly experienced body of talent, familiar with African, European, or Asian culture. Once both sides signed contracts, the international market for the new product was secured with great fanfare. In contrast, Huawei had been established for 10 years, yet had only ventured into international markets for the past three years. What's more, it was losing more battles than it was winning.

In talking with people in the telecom field in America, Ren learned that they generally felt that there would be a revolution in the telecom industry and in networking technology within the next decade. Either advances in computer networking would supplant telecom, or telecom technological advances would incorporate computer networks. Because of this, under the impetus of private capital, a large group of companies were madly putting enormous sums of investment into R&D, to explore and to create new growth opportunities.

All this sparked a wealth of ideas in Ren. In 1997, the R&D budgets for Lucent and Nortel were roughly USD 3 billion, while that of Huawei was a mere RMB 400 million. Huawei could not be remotely compared to these two giants in North American telecom, but also it could scarcely be compared to several tens of thousands of emerging IT firms in Silicon Valley, who were supported by venture capital funds. Where would Huawei be when this new technological revolution exploded?

Ren's trip in 1992 had merely skimmed the surface of America, mainly impacting him through visual impressions. This time, he investigated America at a closer distance. The trip had a powerful effect on his thinking. "Surveying the history of America's IT industry is enough to terrify a person," he later commented. "If you collapsed all of the 500 years of the warring states period into one day, who would be considered the ultimate heroes?"

"Within 10 years, the telecom industry and networking technologies will go through a revolution. Senior management at Huawei already recognizes this. When this revolution arrives, Huawei may not be able to grab hold of the halter of the water buffalo, but it certainly wants at least to grab hold of the tail. That is the only way to become a large international company. What's more, the revolution is almost upon us. Only in the midst of revolutions are new opportunities born."

Clearly, Huawei was not planning to leave the field of action on the eve of this new technological revolution. The only course of action was to burn the bridges behind them and force the troops to fight. Otherwise, as the new-tech revolution erupted, Huawei's entire army would be destroyed. Because of this, not only did Huawei need to revamp itself, "wash its mind" and change its very face, its bones, its lungs, its way of thinking, but also it had to increase investment and choose an American company to serve as its guide.

Ren brought tremendous results home with him from this trip to America to "find the way." He found the way, which was indeed the only possible path to take, the proper response if his small Chinese workshop intended to emerge from its chaotic state.

"The Brave One Wins the Fight on a Narrow Road"

At the annual celebration that Huawei held on January 16, 1998, Ren made a revealing statement: "We are celebrating today, but even as my mouth is open up here, I don't know what to say. The past is behind us and what we will face in the future is uncertain. The difficulties, the dangers, make it hard for us to think that we should let down our guard."

"Huawei is extremely immature as a company, and unfamiliar with global management. Have we all thought about how we are going to survive in the international arena? Faced with the new-tech revolution, soon upon us, will we remain fighters or will we surrender? Will we be victorious or eliminated? History does not let things happen by fluke—we must prepare."

Everyone suddenly went quiet. Ren's deep martial voice echoed throughout the hall. "History provided us with this opportunity, but also with unprecedented problems. As we have evolved from a very small company, we continue to radiate bad habits. As we have expanded, our organizational systems have not expanded in comparable ways. Our management is thin and vulnerable. This will undoubtedly come out as we go through high-speed growth."

"The market has no time to wait for us to grow up. It is not our mother. It has no patience and no mercy. There is no exit for us if we retreat or even if we fall behind."

Finally, in a particularly grave voice, Ren said, "Huawei is in a super-abnormal period of development. The most serious problems facing us are not our adversaries, and also not our human resources, or capital. The biggest enemies we face are ourselves. Whether or not we can conquer ourselves is what ultimately will determine our failure or our success."

"On a narrow road, success goes to the most courageous. We absolutely must break through our psychological barriers, put effort into our management and service, bit by bit, little by little. As the flood of the market challenges us, we must bravely go to face it."

After he returned from his investigation in the United States, Ren's sense of crisis was even more acute. Employees were now infected with his state of mind. On the way home that evening, they debated everything he had said. "The boss isn't happy. He has things on his mind. Do you think operations have run into some problem?" "Every time he goes abroad, he comes back and something big happens. You just wait—there are going to be changes around here."

The Huawei Oath

On March 23, 1998, after three years and eight drafts, the final draft of *Huawei Basic Law* was reviewed and approved by the committee after consolidating the minds and efforts of the six professors and also more than 10,000 Huawei people. This then slowly pulled open the curtain on a new period in the company.

"In order to make Huawei become a global first-class equipment supplier, we will never go into the information-services industry. Since market pressures are unreliable, our internal mechanisms must always be in a state of activation."

As this first sentence floated out over the assembled people in the hall, everyone drew a collective breath. Since *Huawei Basic Law* had already been discussed over three years of constant revisions, everyone was quite familiar with the phrase "global first-class," but this phrase about "never go into" had never appeared before in the discussion drafts. Cadres in the room all thought it highly strange. It then became the subject of intense debate.

Many cadres believed that as a manufacturer of telecom equipment, Huawei was currently in the midst of high-speed growth that must inevitably end. The demand for telecom equipment was limited, which meant that the market would eventually be saturated. The information services industry was instead an industry that could expand indefinitely. Given Huawei's capabilities and its determined spirit, this "service" type of income could even surpass the income that Huawei made from selling equipment. Because of this, many cadres recommended that the phrase "we will never go into" be deleted.

Ren offered his own thinking on the matter. "We are intending to 'break the cauldrons and sink the boats after crossing,' that is, cut off all means of retreat and force ourselves to move forward. Saying that we absolutely will never go into the information services industry is a way of doing that. Through transmitting into our organization the unreliable nature of market forces, our internal mechanisms have to remain in a perpetually 'activated' state. When faced with a fatal situation, we fight to live. Putting us in that position may well force us to become a first-class equipment supplier."

With that, people finally understood the full significance of this phrase. Once Ren had decided on something, given his unyielding character, he would see it through to the end. Once debate finally died down about "never go into," phrases then followed in the bylaws of the company that said such things as "Depending on the demands of sustainable growth, we will set up reasonable profit ratios and reasonable profit targets, that is, we will not focus exclusively on maximizing profit." This phrasing also stirred up debate, but the final draft of *Huawei Basic Law* passed unanimously when the committee put it to vote and all raised their hands. After the final committee meeting ended, however, cadres continued to express their reservations about the growth strategy of "never go into" for quite a long time.

Ren addressed these reservations in an internal meeting. This also was the time he set forth the famous conclusion that "The company cannot put on red dancing shoes."

"Red dancing shoes are highly seductive. They are just like profits in areas outside telecom products. However, once the company puts them on, it can't take them off. It is obliged by them to keep on dancing, all the way up until it dances itself to death."

Around this same time, Ren continued to emphasize what he called the concept of "the endeavor." Huawei's endeavor was to become a world-class corporation. This was the glorious dream and the long-term aspiration, and it was something that employees must keep firmly in mind. He emphasized this on several occasions as various classes of employees graduated from Huawei University, as well as during meetings of the marketing committee and R&D team.

"The information age calls for heroes, and history calls for heroes. Huawei's young people should become the heroes of the world's telecom industry. Who is to say that our PhDs, taught on native soil, won't be tomorrow's heroes around the world? Those Huawei employees who have real ambition should dedicate themselves to the glorious mission of mankind's telecom technology."

In addition to pep talks within the company for employees, Ren spoke on the same subject on June 19, 1998, and June 24, 1998, at meetings outside the company. At the first event, he delivered a report to the China Telecom Research Team; at the second, a forum for senior management at the headquarters of Unicom, he delivered a report called "How Long Will the Red Flag of Huawei Actually Be Able to Fly?" At both of these, he gravely told his Chinese listeners about "Huawei's oath to become a world-class company."

"People may think it simply amusing that such a tiny company as Huawei should propose such a crazy slogan. But it was precisely such goals that led us to where we are today. The gap is still huge, but we are reducing it every year. If Huawei does not intend to die out, it has to establish the concept of being 'world class' in the minds of both employees and customers. We must establish a direction and goals for the company if we are to generate a sense of trust and reliance among our customers. We also must make our employees feel that they are walking on solid ground even as they aim for grand objectives."

Ren's ambition undoubtedly stirred many people. Back in October 1994, at the first Beijing International Telecom Exhibition, he declared the intent to "divide up the world in three parts, Huawei, Siemens, and Alcatel." At that time, Huawei was not even seven years old. It had not been in the market for program-control switches for even six months. Many competitors scoffed at his ambitions. In the four years to come, however, all the foreign "princes" who had long monopolized Chinese turf gradually withdrew from the China market. Huawei, indisputably, continued to rise. Who was now going to doubt that Huawei would one day truly be on a par with Siemens or Alcatel in international markets?

Going from "Rule by Man" to "Rule of Law"

Since Huawei had now stepped onto a path that could take it forward, the question became how to use institutionalized systems to consolidate, constrain, and also lead Huawei in the right direction—how to turn it into a world-class corporation according to the direction the company had decided to take. This constituted a major turning point in Ren's professional life.

Prior to concluding the final draft of *Huawei Basic Law*, Ren had told the drafting group many times that in its first founding, Huawei had mainly relied upon entrepreneurial intuition and gut instincts to seize opportunities. The ability to lead troops and get them to fight their way forward is what had enabled Huawei to go from being a small workshop to an enterprise of at least moderate size. After 10 years of growth, the question now became what to hang onto and what to discard. Given critical analysis, could the company carry its legacy forward and still open up future prospects? This had to be the focus for the second founding. Looking out over corporate histories in general, many companies in China had failed after reaching a certain size, whereas western companies at this same stage had gone on to prosper. The thing that made the difference was rule of law. For Huawei, the key factor now became the ability to go from rule by man to rule of law.

"The most important feature of Huawei's second founding is that it must reject the authority of 'rule by man.' The core issue is to move in the direction of being regulated and institutionalized, to dampen down the influence that I myself have on the company. It is to change the way one individual calls the shots, lays down the law. The core issue is to keep the company from ceasing to function after the entrepreneur leaves."

"By muting the colors of the founder's influence, strengthening professional management, setting up scientific decision-making mechanisms and management mechanisms, we thereby gradually get away from reliance on technology, human talent, and capital. Instead, we realize 'rule by not-doing.'"[2]

Ren made these determinations after close study of Asian and western corporations and case studies that involved both failures and successes. Based on his conclusions, the drafting group then went a step further in formulating a set of subsidiary laws that were under the part in *Huawei Basic Law* that talked about "decision-making systems." One example was the management law of the Huawei Corporate Committee. This said that all major matters of the company had to undergo collective policy making by the committee. This was to mute the individual hues of the entrepreneur, to reduce the influence of an authority figure, and to avoid and reduce mistakes that might be made by individual determinations. Strengthening collective leadership was done to raise both the quality of and the scientific nature of decision making.

The birth of this new decision-making system made it clear that Ren had already distanced himself from the autocratic manner by which founders of Chinese corporations typically consolidate their ability to lay down the law as individuals. It also expressed the fact that Huawei had taken the critical first step in moving from decision making by an individual leader to collective decision making, from rule by man to rule of law. In internal meetings, Ren also said repeatedly that "muting the colors of the entrepreneur" did not mean dampening the spirit of entrepreneurship. The second founding of Huawei was a dilemma that involved broader issues. In 5,000 years, China had never produced a large corporation on the order of Hewlett-Packard or IBM. Relying just on the efforts of one or two generations of people was not going to be enough—this was going to take several generations' worth of unstinting effort. Because of this, the successors of Huawei had to turn such things as an entrepreneur's force of personality, ability to lead, and individual propulsive energy into the intellectual assets of the company at large. They had to turn the

---

[2]This takes off on the Daoist philosophy of action through inaction, or governing through not governing: all things are accomplished by not taking action, *wuwei er wu buwei ge*.

entrepreneur's understanding of life itself into the kind of soul and spirit that could be transmitted generation after generation.

In March 1998, when *Huawei Basic Law* was officially inaugurated, the six professors of Renmin University who had constituted the drafting group described the process in a book called *Order Out of Chaos*. This described how the idea began, was incubated, and came to fruition. The book was published and distributed by China Telecom Publishing House. At the request of the drafting group, Ren wrote a Preface for it titled, "From the Realm of Necessity to the Realm of Freedom." This described the hopes that he placed on *Huawei Basic Law*, and it set forth his ultimate mission:

> *The key to whether or not a company can govern itself peacefully over a long time depends on whether or not its core value system is adopted and accepted by each succeeding administration. It depends on whether or not our many thousands of employees agree with it and, indeed, are willing to fight to put it into effect. The* Huawei Basic Law *has already clarified our outlook on core values. The hope now is that our many thousands of employees will accept this outlook and will put it into effect in actual practice.*
>
> *Our outstanding employees must constantly ask themselves, generation after generation, whether or not the internal and external laws that determine growth are understood clearly. Does this allow management to achieve a state of "rule by not-doing"? If we work at this, we should definitely be able to go from being a company that operates in the realm of necessity to being one that operates in the realm of freedom.*

Huawei was clearly facing the following trajectory if it was to go from being a "China-style small workshop operation" to becoming a multinational. With *Huawei Basic Law*, it was understood in no uncertain terms that the company had to align its management with international practices. It had to create global processes and professional teams that could then truly operate on their own without "rule by man." Its mission had to be unified, along with its ideologies, ambitions, concerted efforts, and ability to transmit the spirit of entrepreneurship.

Huawei Basic Law established a precedent for companies in China. Not only was it a constitution for an enterprise, but also it led the way in going from a rule by man to a rule of law for the country in general. At a point when Huawei was inexorably rising in the China telecom market, under the "iron wrist" of Ren, the company now launched a movement to reform management. Over time, it was to spend RMB 2 billion on aligning management with international practices.

CHAPTER SEVEN

SETTING UP A MODERN, AMERICAN-STYLE MANAGEMENT SYSTEM

Chinese people are far too "clever," which is why they have remained poor for 5,000 years. Japanese and Germans are not all that clever, yet they are many times more prosperous than Chinese. If Chinese do not "regulate" their cleverness, they will find themselves tricked by it.

—REN ZHENGFEI

"It doesn't matter if the steps are big or small—both move forward the progress of mankind." This IBM ad, in a bustling part of Beijing, epitomized the aspirations and creative spirit of Big Blue.

Seventy IBM consultants strode into Huawei's headquarters in Shenzhen in the latter part of August 1998. They settled into Huawei's headquarters, which at the time was still located on Kefa Road in Shenzhen's Science Park.

Ten years later, on the evening of February 29, 2008, the CEO of Huawei, Sun Yafang, officiated at the final send-off of IBM consultants. In a ceremony that included some 50 of Huawei's senior managers, she thanked more than 150 IBM consultants for the guidance and assistance they had given Huawei over the years. IBM and Huawei had worked so closely together, "shoulder to shoulder in the battle," that one of the vice presidents responsible for management reform at the company became quite emotional. "This may be just a consulting project for IBM," he said, "but for Huawei, this means the birth of a new life."

The IBM consultants were equally moved. "We put 10 painstaking years of our life into this effort," said one named Arleta. "We put our very souls into Huawei. We feel extremely gratified and proud to have contributed to turning Huawei into a multinational company."

Not long after, the famous consulting group called Informa, revealed in a survey, published in *Global Mobile,* that Huawei had earned USD 18.3 billion in operating revenues in 2008. That meant it had surpassed Alcatel, Lucent, Nortel, and Motorola and was now sharing the entire global telecom market with Ericsson and Nokia. At the same time, *Businessweek* voted Huawei into the world's top 10 most influential companies. Huawei was now put in the same ranks as Apple, Google, Unilever International, Walmart, and Toyota. One year later, data published by *Fortune* magazine showed that Huawei earned USD 21.8 billion in 2009 in operating revenue and thereby entered the ranks of the Fortune 500.[1]

To understand how Ren Zhengfei cast off his sandals and put on American shoes, however, we have to shift our focus back to 1995. That is when his troops began to march in unison, ultimately beating out the well-armed troops of many other countries.

Why Go for Wholesale Westernization?

Each person in the core leadership of Huawei in the early days had some expertise or other. Ren's expertise lay in installing heating systems. He had no experience in management whatsoever, since his training and his period in the army all involved technical subjects. Sun Yafang's expertise was in electronics. Zheng Baoyong and Li Yinan had studied photoelectric applications. What's more, all of these people had graduated from college fairly recently. Without exception, all other members of the core leadership were engineers, and most had been involved in research and development. Huawei had to grope its way forward under the leadership of a group of people who had no experience in running a large high-tech company. Once the company survived a very tough period in the early days, it was unavoidable that the lack of management expertise

---

[1] In Chinese, the Fortune 500 is known as the World 500. Thus we use that term in the rest of this book.

should cause the company to take wrong turns and pay a high price for its mistakes.

As one example, everyone in the R&D team was young, adventuresome, and willing to take risks. Based on an acute "sense of smell" and abnormally high intelligence, they managed to develop fairly high-level telecom products in a very short time. However, youth also had its fatal weaknesses. Over the long term, Huawei lacked a rigorous and well-defined plan for R&D. Basically, the team would quickly organize an R&D subteam to pursue whatever product line was selling well on the market. These people would work night and day and eventually come up with similar products. This method of R&D was results-oriented, however. As a result, Huawei had twice as much waste in its R&D investment as the rest of the world, and its R&D personnel focused on core technologies while slighting the stability and reliability of products. A great number of unready products were pushed onto the market, which eroded the trust of customers that the company had worked so hard to cultivate.

Ren once noted painfully in a meeting that Huawei had to spend RMB 500 million to replace more than 1 million customer motherboards, since the design of Huawei's control-process switches had problems when used with them. Optical transmission and VPN systems also had various problems due to the R&D process, leading to constant equipment "accidents" once products were put out on the market. Returned goods from customers then created another RMB 1+ billion in losses.

What's more, given fierce competition, sales staff made all kinds of promises to customers in order to get orders. After signing contracts, not only did they then ask the R&D staff to design changes, but they also asked production departments to shorten production time. Design was then changed at random, leading to both uncertainties in costs and an inability to guarantee product quality. Once quality problems became a frequent occurrence, large numbers of service people had to fly back and forth across China, trying to repair things. Profits were used up just on plane tickets and hotel bills. What were they trying to repair? It might only have been a trivial part of the production process. But these tiny mistakes forced Huawei to pay a price that was thousands, indeed tens of thousands, of times the original problem. Due to the culture that Ren espoused, the "cot culture," and to the ethic of hard work,

Huawei people sacrificed any resting time and tried to make up the wasted money by paying for it with their own time. This did not solve the essential problem of the company's inability to create workable systems in the first place.

In July 1995, a visitor to the company asked Ren to describe what he felt was the greatest lesson he had learned in founding Huawei. Ren thought carefully before responding: "The waste has been considerable. The lessons have cost us several billion RMB. Several billion RMB have simply been scrapped." The friend tried to comfort him, saying that Huawei was still young. It needed to carry on and redouble its efforts. To this, Ren responded bitterly, "Youth is no excuse for us. We can't use it to forgive ourselves. After all, Microsoft is just as young—why could they do what we could not?"

He drove the point home. "In 5,000 years, China has not been able to produce a large company such as IBM, Lucent, Hewlett-Packard, or Microsoft. Our management systems, management rules and procedures, the caliber of management, the quality of our psychology, not to mention our technology, are inadequate to support the creation of such a company within China. All we can do is rely on ourselves to make progress, since otherwise there is simply no hope. But this kind of groping forward, feeling for the stones as you cross the river, is unimaginably difficult."

After massive amounts of waste and quality control problems, Ren declared, "We must not forget what has happened in the past, since we need to use that to guide us in what is still to come." The company gradually eradicated some of the worst management problems that universally plagued China's "small workshops," but with considerable difficulty. Never for a moment has Ren relaxed his efforts to go much further.

Regulating the Behavior of Clever People

In the three years between 1992 and 1995, Ren led delegations to the United States, Germany, and Japan to conduct investigations. He was surprised to find that Huawei's disparity with world-class corporations lay not only in cutting-edge core technology, standardized management, and standardized production, but also in the level of professionalism of employees and in the psychological quality or "caliber" of employees.

The professionalism of Bell Labs impressed him in particular. Scientists there did not rest on their laurels when they achieved something. They did not consider it "enough," but kept on working on the next thing, meanwhile scribbling formulas on any available surface—their shirts, the walls—and continuing to make outstanding contributions to humankind. He made the comment, "The spirit to battle onward is definitely not unique to our communist party members! America has it too! And it is precisely this kind of unselfish effort that has created America's prosperity."

When visiting Siemens in Germany, Ren noted that quality control personnel used specially designed magnifying devices to test each circuit board on the production line from six different angles. When the process required several hundred workers to do a particular action, they did it in precisely the same repetitive way. He commented that West Germany's economy would never have recovered as fast as it did after the Second World War without that spirit of impeccable precision.

In visiting Japan, Ren happened to pass by a construction site near Tokyo. Watching from a distance, he saw a number of robots doing the same repetitive motion. Only when he came closer did he discover that these "robots" were in fact very ordinary construction workers.

Another time, he saw an elderly person walking back and forth on a slope, looking for something. When he approached, he asked the old man what he was doing. The gentleman said he was using a sharp-pointed stick to pick up bits of paper in the grass.

As with Germany, his thoughts connected human diligence with economic prosperity. Japan is an island, small and narrow, he later noted.

Not only does it not have coal or iron ore, but it also has no oil. Yet the Japanese work ethic enabled the country to rise up in just two to three decades after World War II, to become a strong world economy. Not only do the Japanese have the ability to endure, but they have a very respectful approach to things.

In contrast, Huawei's many years of enormous waste, its losses, have not in fact been regarded as a major problem by our cadres.

Nobody loses any sleep because of it or puts any heartfelt effort into changing it. The rate at which we make mistakes in our contracts remains high. Everyone raves about America's innovative spirit, Germany's drive, and Japanese meticulous management, but we ourselves do not pull together and take action.

Many of the whole series of problems that we have been experiencing relate to the caliber of management, the degree of "cultivation" of our cadres. And often that relates not so much to skills training as to a sense of responsibility and a sense of mission. We still, for example, have too few employees who are willing to stiffen their spines and stand up for things. Negative attitudes are too common among our managers. Our production systems are not at a point where management can go step by step in forcing everyone to adopt more reasonable procedures.

After these overseas investigations, Ren did a close comparison between Huawei people and the innovative approach of Americans, the industrious attitude of Germans, and the respect of Japanese for their work. He described his conclusions in an internal meeting as follows. He said that it was on account of his conclusions that, on one hand, he was mobilizing mass movements and launching a self-criticism process in order to upgrade the caliber of thinking of employees and their sense of potential crises. On the other hand, he was actively pushing forward three standards by which products and marketing should be evaluated: "the level of advanced technology in America, the high degree of stability of products in Germany, and the low production cost of products in Japan." At the same time, he asked senior managers to squeeze out time to go abroad to look things over for themselves. In every respect, they should seek to find ways to shorten the management gap between Huawei and the world's more famous companies.

Choosing a World-Class Teacher

In mid-1995, when Huawei's standing vice president, Zheng Baoyong, led a delegation to Shanghai to visit the Alcatel-Shanghai Bell company, he found that this joint venture, in existence for less than 10 years, had been able to come into alignment with international standards quickly by importing overseas advanced management systems. This was an exciting realization. After returning to the company, he delivered his findings at a meeting and recommended importing MRP II (Manufacturing Resources Planning) to establish a full set of standardized business procedures and management systems. Ren gave him full support. After looking into it, Huawei invested more than USD 10 million in importing two sets of advanced corporate management systems from Oracle in the United States and SAP in Germany.

However, once the American consultant and the German consultant came to Huawei, each had his own ideas about how to implement these two systems. Each, according to his own opinion, "localized" or "Sinified" the programs. Although business procedure systems were ultimately established in Huawei after being formally adopted in April 1997, neither system achieved the desired results.

Huawei achieved two strategic transitions in the three years that it took to draft *Huawei Basic Law*, between 1996 and 1998. One involved "using the countryside to encircle the cities," and one was "setting sail for overseas markets." After three years of bloodbath-type battles, Huawei's new forces came to the fore in the domestic market. They were successful in sweeping away all obstacles. In the overseas market, however, the company suffered one defeat after another. What's more, as China's fixed-line telephone market gradually became saturated, opening overseas markets became the company's sole option. If Huawei was to win a decisive battle overseas against "well-armed proper armies," it must turn itself into a "proper army." That is why it spent an enormous amount of money to import advanced management systems from Europe and America. Making Huawei into a world-class enterprise with a "pure bloodline" now became the main issue that Ren urgently felt the need to resolve.

To do this, starting in 1996, Huawei brought in the Hay Management Consulting firm from the United States. This company set up a human resource management system that centered on performance evaluations and salary systems. However, if Huawei's three main management centers of R&D, production, and supply were not coordinated with this, and did not carry out corresponding reforms, Huawei's human resource management systems would simply stay at a primitive level.

In an internal meeting, Ren therefore announced in explicit terms that the company had to go further along in this process. "Huawei has not only grown but it has grown in the context of China. That means it has had to rely on itself and its subjective intuition, since external resources for companies are nothing like as abundant as they are in the United States. That process has lacked a rational, scientific approach and standardized procedures. Because of this, we now need to take advantage of the experience and methodology of the United States. We must fully understand the way that western companies think about salaries and compensation, things that the Hay company has provided. We cannot simply, mechanistically, import bits and pieces, fragments that we like best."

The most urgent concern at this point was to set up a modern human resources system and then, on that basis, to align the company's R&D, production, and supply chain with international practices. There were all kinds of models that might serve this second part of the process, but none seemed right for wholesale revamping of the corporation. Ren leaned in favor of first one consultant, then another, but it was only in December 1997 that he finally found a "teacher" that he felt could help the company "change the very marrow of its bones."

In October 1997, China and the United States issued a joint declaration, announcing that the two countries intended to strengthen cooperation. They would make all efforts to set up a strategic partnership that was aimed at the twenty-first century. This opened a new chapter in the development of the relationship between the two countries. In the week before Christmas in 1997, Ren made another trip to the United States and examined world-class enterprises up close, including Hughes Network Systems Lucent, and Hewlett-Packard, but he focused on a visit to IBM as being most representative of "American" enterprises.

As a leader in the IT industry, IBM not only provided equipment and overall IT solutions to global markets, but also was moving into management and consulting services as a new growth area. Huawei's visit was warmly welcomed since the company was looked upon as a key customer in this regard. Even though it was Christmas eve, and most large American companies had closed for the holidays, IBM's CEO and other senior management were working as usual, and they systematically introduced their management practices to Ren.

Handing Over RMB 2 Billion in Tuition to Learn from IBM

IBM's history of going from making time clocks and tabulating machines to becoming a peerless giant with USD 90 billion in annual revenues took 100 years of hard work. Since it enjoyed success for so long, however, its organization became bloated and bureaucratic to the point that the company was too slow to seize new opportunities in an era of personal computers and networking technologies. The market for mainframe computers, upon which it heavily relied, was being battered to the point of being obliterated. In 1992, IBM ran into the toughest financial situation it had experienced so far. Sales revenue stopped growing at all, and profits fell dramatically. Losses of USD 16 billion were leading the company in the direction of bankruptcy, and media reports already were declaring that the company had one foot in the grave.

In early 1993, the first CEO not to be hired from within IBM, Louis Gerstner, was given the task of rescuing the company. After thorough analysis, he took forceful action. He eliminated redundant personnel and cut the huge bureaucratic system which was "for people and not for getting things done." He set up incentive policies and mechanisms for defining production targets. He abolished the old corporate culture, which had become rigid, and set up a new corporate culture "that was customer driven." Meanwhile, with respect to IBM's technology which was powerful but slow-reacting, Gerstner courageously adopted the R&D management model called Integrated Product Development (IPD). He both shortened the time it took to get products to market and raised the profitability of products by reorganizing both procedures and products. Finally, he was able to transform IBM from being technology driven to being market driven.

After five years of rousing the company to better performance, and after paying the heavy price of more than USD 8 billion in administrative costs and reduction in forces of 150,000 personnel, IBM was finally resurrected. In 1997, its share price went up by four times, and its sales reached USD 75 billion. In addition to glorious results, it wrote what could be called a legend in modern corporate history.

During Ren's visit, senior managers conducted a detailed explanation of IBM's operations for an entire day. They went from pre-research on products to project management, and from production issues to the conclusion of project life cycles. To allow Ren to become more familiar with IPD, an IBM vice president presented him with a work published by Harvard University on the management of R&D. Ren later learned that Hewlett-Packard and Lucent were also implementing similar R&D models.

After this day of learning about management, Ren had a new appreciation for IBM's effective management and quick responses, and he had a new understanding of Huawei's own shortcomings and how to resolve management deficiencies during a period of fast expansion. He began to see how to deal with low efficiency and severe waste, and he had new ideas about how to make fewer wrong turns in the future.

One thing that most impressed Ren was Gerstner's personal style as a manager. Not only did he have a rare determination and the courage to make major changes, but also he was extremely low key. In the nine years that he had served as IBM's chief executive officer, he rarely gave interviews to the press and rarely appeared in public. It was this matter-of-fact approach to getting things done that enabled him to concentrate all his efforts on setting up the kind of corporate culture in IBM that focused on first-rate business processes, highly transparent strategic development, and high results. It was this that ultimately brought IBM back from the brink of disaster and took it to another peak.

All this excited Ren. If Huawei wanted to be as powerful as IBM, not only should it take Gerstner as a role model but also it should ask IBM to serve as "teacher." No matter what the cost, it should try to transplant IBM's management expertise over onto Huawei's systems. This seemed to be a necessary path for Huawei if it intended to be world-class, and if it intended to become more regulated, professional, and international.

If the company cuts its size too much, it becomes less competitive. If it expands too much, it cannot have effective management and again runs the risk of failing. Only by strengthening management and services can a company create the basis for its ongoing existence.

In America, every family puts lights everywhere at Christmastime. We, however, were staying in a small hotel in Silicon Valley and turning on no more than the wall switches to our rooms. We did not go out for three days as we held a meeting to digest all that we had seen and heard, in order to take it back to China and pass on the report. The price that IBM had paid for its own experience, including the direct cost of several billions of U.S. dollars, was a priceless treasure to us.

A Hero Who Does Not Fade as a Short-Lived Flower

In February 1998, after Ren returned to China, the newspaper *Huawei People* published an article called "What We Learned from the Americans." This described in detail Ren's thoughts, concerns, and astonishment at what he had learned. He asked the editor to print a photograph next of Andrew Grove, chairman of Intel, next to the article, to spark readers' interest in America's innovative mechanisms, its spirit of getting things done, taking risks, fighting for something. He also put in the following caption: "America's high-tech field takes in highly qualified immigrants from around the world. It provides an atmosphere that allows for innovation and also failure, and which focuses on start-ups and then forces them to be tested by the market. Its most famous 'doer' is Andrew Grove, who personally is an expression of all of this. We hope that Huawei can also produce 'doers' like Andrew Grove."

Over the next three months, after *Huawei Basic Law* was issued, Ren began to lay the foundation for a whole series of management changes.

At one management meeting, he gave an address that he titled "Not Being a Flash-in-the-Pan Hero," in which he had the following to say:

We have spent 10 years achieving hard-won successes, and we may indeed have more and more things to brag about both inside China and overseas. Will we become complacent as a result? Will we perhaps develop unhealthy habits? Will Huawei people begin to wear a kind of smug arrogance on their faces? I wonder what they will base their arrogance on, since in fact we do not know what will sustain our success. We do not in fact know at this time if the path before us is going to broaden out or become narrower and tougher.

Some people ask when we are ever going to be able to take a break, draw a breath. I tell them only when the nails are hammered into your coffin can you relax. The only thing that doesn't change is change itself. What that means is that management must eternally keep improving.

Ren went on to say that management was a perpetual issue for businesses everywhere, that is, a perpetual problem. Huawei intended to become an international major company—everyone was already in agreement on that and doing "thought preparation" and organizational preparation. In terms of management ways and means, however, it lacked experience. In the second start-up of Huawei, management was the thing that people did not see because it was invisible, yet it was the decisive issue. It was the thing that would determine Huawei's very ability to survive.

Once American companies win technological breakthroughs with their products, they can wave the banner of their products and hire talent from anywhere in the world and break into global markets. When Huawei makes breakthroughs, this is not the case. Not only can it not break into the rest of the world, but it may not have that much of an advantage on its very doorstep.

Because of that, Huawei as a company must be absolutely determined to improve its management and services. The moment a new opportunity arises, we must grab hold of it and we may then become a giant. Right now, we have opportunities but we cannot grab onto them. At the most, even within China, all we can do is take over some nonmainstream markets. Meanwhile, we are conceding the great majority of international markets to western companies.

Finally, Ren said with a certain resonance:

*The lack of good management methods will lead to low effi-
ciency, which in turn leads to death. Western companies have at
least three times the per-person results that Huawei has. What
are we wasting in this process? We are wasting both resources
and time, and this waste is being caused by Huawei's ineffective management. If we do not have any ambition or intent to
improve management, in point of fact the company can already
be considered dead.*

*Whether or not Huawei goes under is completely up to us. It
is determined by whether or not we improve management. We
hope not to be a hero that blooms just briefly in the sun. Huawei
has accomplished a few things, but if we rest on our laurels, the
bloom will fade. Will the flower bloom again? That is question-
able. In the information industry, the minute you fall behind it
is hard to catch up again.*

In August 1998, Ren again called together senior personnel for a
management meeting, including close to one hundred vice presidents
and senior-ranked cadres. He announced that a project with IBM was
formally now launched, called "IT Tactics and Plans." This plan incorporated eight different changes in management over the next three to
five years that were necessary if Huawei was to transform itself into a
world-class corporation. These eight included IPD (integrated product
development), ISC (integrated supply chain), IT systems reconfiguration,
and four key aspects of budgeting.

After this meeting, Ren instructed staff to prepare a number of new
rooms near the beach at Huawei's already crowded corporate headquarters. Furniture was brought in and the offices were redecorated according
to IBM's requirements, so as to make the guests feel at home. In addition, per IBM's prices, depending on rank, the seven consultants were to
be paid between USD 300 and USD 680 per hour. Clearly, completing
this five-year management change was going to require an investment of
RMB 2 billion, just in tuition.

As the first group of IBM consultants started work, this move to revamp Huawei's management was officially inaugurated. It was to proceed under the "iron wrist" of Ren.[2]

Big Blue's American-Model "Clinic"

As noted above, Huawei began to import various overseas management systems in 1995. For assorted reasons, none of these came up to expectations. Because of this, quite a few cadres were concerned that this new effort, costing RMB 2 billion, would be the same. As the work unfolded, however, these cadres discovered that the IBM consultants were totally different from those who came before. Not only did they have profound respect for their work, but also they were highly principled.

In interviews and discussions, the consultants asked Huawei cadres to speak English when they raised questions about various issues. To unearth firsthand material, they also spent a great deal of time discussing Huawei's management issues directly with cadres. After more than one and one-half months of interviews, they had come up with the main threads of Huawei's development history and, by "reverse thinking," they were able to dissect the overall state of affairs regarding Huawei's management.

After personal observation of the work ethic, the discipline, and the conscientious nature of the experts, cadres were won over. They were willing to submit to the professionalism of IBM's consultants. On September 20, 1998, Ren led a group of several dozens of senior Huawei personnel to a hall where they arrived far ahead of time. There were soon no empty seats in the audience. When the proceedings began, IBM consultants stood before the podium and went through "10 Major American-Style" diagnoses that they had come up with, systematically and in detail.

First, the company lacked an accurate way of forecasting customer demand. This led to useless products and massive waste of both resources and investment costs.

Second, the company did not have procedures that crossed departmental lines. Each department had its own procedures, and any interface

---

[2]This is the Chinese way of saying that someone rules with an iron fist. The phrase in Chinese is quite descriptive, and the author keeps it intentionally.

with other departments relied on artificial links so that there were lapses in operating procedures. Each department was cut off from the others.

Third, in organizational terms, "self-interested departmentalism" was pronounced. Departments erected high walls around themselves so that they could run their own affairs. This led to increased internal expenditures.

Fourth, people had insufficient technical skills, and work was not standardized so that results relied on individual "heroes," but it was hard to duplicate the success of these heroes.

Fifth, project planning was ineffective and was even implemented in a "chaotic way," with no control over modifications during the rampant issuing of "new versions."

Once the consultants finished describing the fifth consideration, a commotion broke out in the hall.

Prior to this time, Ren and other senior managers had been quite clear that there were management problems in the organization, but they had no way to describe these accurately, to say exactly where the illness resided. The problems that the consultants laid out were quite acute. They touched upon very sensitive parts in Huawei. Ren's expression now became extremely grave. At a break in the meeting, he asked the manager of the IT department to tell cadres of relevant other departments to come to the hall. More than 50 additional cadres gradually filed in. Since the room was already full, and there was no way to change halls on such short notice, Ren indicated that everyone should move tables and chairs to one side. In the space that was created in the middle of the room, he then had all the cadres sit on the floor.

A question-and-answer session ensued, after the Report on the 10 Major Diagnoses. In this, both Ren and Li Yinan questioned why the report had positioned Huawei as a "production-quantity type of company," when according to Huawei's policies, it put 10 percent of sales into R&D every year. Shouldn't it therefore be considered an innovative-type company? The consultants not only did not answer the question on its face value, but they rejected the question out of hand, and in strongly worded terms.

After the meeting ended, Ren was pleased and congratulatory. "They dared to go against what we ourselves believed. We were right to ask

IBM to be our teachers. This is precisely the kind of consultant that Huawei needs."

Hearing this, senior managers thought back to a previous conversation Ren had with the consultant from the Hay company. At that time, he also invited a number of cadres in to listen. He had said quite sincerely to Ms. Wei, the consultant, "I am delighted to be challenged by Ms. Wei today. I will be very glad to answer all questions she puts to me—it is a rare opportunity to have an international examiner grill me in this way, and I hope that I come through all right. I'll be as good a student as I can."

These words were in fact for the benefit of the other cadres in the room. He was telling them that they should have a conscientious and selfless attitude to learning new things. Today, in using the same technique, Ren was telling all the assembled students that they must also have a humble and willing attitude.[3] Things that he could not have anticipated were now shortly to follow, however, one after another.

We Absolutely Must Rectify our Attitude Toward Learning

The lessons and diagnoses of the IBM consultants drew blood. The next three years—1996, 1997, and 1998—brought multiple increases in sales revenue to Huawei, so the admonition to be selfless and patient students fell by the wayside as Huawei became accustomed to success, however short-term it might be.

In the stage of training that followed IBM's presentation, created to inculcate certain concepts into the company, many employees actually went to sleep there at their desks. Other senior cadres found excuses to be late or to leave early. Some employees refused to understand just what this thing called "integrated product development" meant and asked the consultants all kinds of questions, including whether it was really suitable for Huawei. At times, they told the consultants point blank that they felt Huawei's procedures were more advanced than those at IBM.

Huawei had achieved several successes on its own doorstep, within China. This kept its employees from "cooling off" and approaching new

---

[3]The term used means pious or devout, in the Buddhist sense.

techniques and lessons with a humble attitude. It was at this point that Ren's "iron wrist" approach was to play the decisive role in the outcome.

On April 17, 1999, at a meeting to mobilize employees to use IPD, he said the following:

Right now, how many of you have any new ideas about how to overtake IBM? If you do, please raise your hand—don't be shy. If you can also generate more than USD 90 billion in asset value, we will definitely ask for your advice. We will no longer learn from IBM. In fact, however, it is apparent you do not have new ideas. Not only that, but you are not studying hard enough, and you are saying things before you understand anything completely. If you want so much to be in the limelight, all I can do is to expel you from the group that is undertaking this project.

Yes, there are plenty of excellent management theories in the world, but we cannot use them all. We would simply become an idiot. If this side manages one way and the other side manages another way, the sum of them would cancel out to zero. We are therefore studying only from one consultant, and we are studying just one model.

What are we to do if someone does not study well? My feeling is we fire him. There is nothing to be afraid of in that—you are all young, you can go back and become engineers. It's not much good being a worker, after all—who wants to lower himself to that?!

On that note, Ren said firmly:

IBM is a truly outstanding company in the world. With great trouble, we have been able to get them to serve as our teacher, and they helped us enormously last year in improving our management. Not only is the way they present themselves very high-caliber, but the methods they are teaching us are things we can actually make use of. We are fortunate in having found a good teacher, and we must make every effort to rectify our attitude about learning from that teacher.

I am extremely moved every time I listen to the lectures of our highly conscientious teachers. Chen Qingru [the IBM consultant] puts so much into this that I am very impressed. If I cannot be similarly impressed by the caliber of the students, then as I see it, the only thing to do is have those students leave their current positions. In these three years, as we go through management transition, it is only natural that some high-level managers become regular cadres. Some regular cadres will become senior managers, and this too is normal. What would be abnormal is if everyone advanced at the same pace.

After the meeting was over, Ren called a meeting of the "leading small group" spearheading the management reforms. He expressed his own attitude on how to respond in a conscientious manner to IBM's teaching. "We must be more willing to accept the consultants' opinions, learn from the more advanced experience of the other side if we intend to figure out exactly what IBM did to become what it is. First, that means senior and mid-level management should really absorb the training and understand it. Prior to understanding, they must not mislead the consultant or we will simply be spinning a cocoon around ourselves. Right now, we are at the stage of beginning to understand the concept of 'IT' in name alone. We still can't comprehend all that it contains. Before we do, we absolutely must not try to change other people's thinking."

Ren's tough approach was quite effective in generating a fear of consequences in cadres and employees. Over the next six months, most employees submitted to learning IPD concepts with a fairly humble attitude. However, once the second stage of the program began to be promoted and implemented, in October 1999, problems again began to appear.

"Putting on American Shoes"

In the past, Huawei often wasted development resources, and also generated massive quantities of unsalable goods in inventory, because it became aware of market needs too late in the development process. This led to the need to revamp design plans or completely redesign products. It led to constant changing of product configurations. New versions were constantly being produced, and development plans were constantly postponed.

In contrast to this ex post facto way of doing things, the core concept of IPD was that of customer demand. Customer needs had to guide the entire process. IPD emphasized that product innovations had to be tethered to market demand and to changes in competitiveness. Given this, the R&D departments no longer had independent decision-making authority over products. Instead, representatives from a number of departments together composed a team known as IPMT (product R&D team), including the R&D, marketing, finance, purchasing, customer service, and production departments. The main task of this group was to figure out the correct orientation for R&D based on customer demand and needs. Moreover, its task included exercising overall regulatory control over the R&D process in order to meet this goal.

IPD truly hit at the core problem in Huawei. Nevertheless, R&D staff in Huawei overwhelmingly felt that the situation in the United States was different, which was why Americans could espouse IPD with such fervor. American corporations were "wave pushers" in the information age. They created the opportunities and then could guide global consumption. They therefore were not subject to the problem of the time constraints of R&D. Huawei was different. As a technology follower, all it could do was find opportunities and seize them after new technologies had already come out. This was a fundamentally different approach to R&D than that in the United States. What's more, Huawei had absolutely no advantage to speak of when it came to technology. If Huawei was even more delayed in its R&D, by figuring out markets first and then trying to cater to them, it would lose its innate low-cost advantage. It would then face all kinds of problems.

This concern was not limited to people in the R&D department. Sales too was worried. After people in the sales department obtained demand information from customers, they would report this up to the R&D headquarters. After roughly three to six months, products could finally be promoted to this specific market. Now, under IPMT, if sales staff were to report on customer demand, it would go to not only R&D but also the other departments. Not only did it have to go through several months of evaluation that included return on investment, a forecast of profitability, a product life cycle, and so on, and that included dozens of evaluations, but also it would have to go through a period of simulated operations and risk analysis. Only then could the determination be made whether to invest money in the project. If you added on a minimum of six months of R&D time, what originally might have been a few months before a product hit the market now turned into more than a year. That included the whole cycle from project development, evaluation, R&D, and finally market promotion.

Not only did the R&D department have doubts about the IPD model, and the sales staff, but marketing as well. Once the second stage of the process of simulated implementation was more deeply understood, the great majority of marketing personnel confirmed that the IPD model would have a strongly negative effect on Huawei's sales results. A number of unfavorable reactions began to be voiced.

"If we aim for such large-scale interdepartmental cooperation, we might as well pull up stakes and go home the moment a project fails."

Faced with this rejection from both R&D and marketing departments, on November 16, 1999, at the conclusion of the report on the first stage of IPD, Ren conceded that Huawei should continue to keep in mind its dream of creating a Chinese model and its own Huawei model. "Still," he said, "we must consolidate what is being imported from outside, improve upon it, and try to consolidate it. In the coming two to three years, it will be important for us to absorb and digest what we have learned. That will be our primary task. After two to three years, we will allow for the appropriate changes."

He also said, however, "IPD relates to the company's future existence and ongoing growth. Each level of the organization and each department must be fully aware of and understand its importance. We first buy a pair of American shoes. It they don't fit our feet, we cut our feet to fit the shoes."

In the meetings of the task force that followed this, Ren noted even more severely, "Size 37 is size 37. If our feet are a little too big, we cut them down a bit. Those people who aren't willing to cut their feet, well, you can go home and grow vegetables. Stand aside."

"We must have a resolute attitude when it comes to pushing forward procedures. Those who can't get used to this must leave. Those who oppose it must be let go. IPD will be put in place, layer by layer [in the organization]. If we don't get this done, I am going to take up a knife and operate on you myself, and that is no joke!"

The Hardest Thing to Change in the World is Your Own Life

Under the impetus of Ren's iron wrist, the IBM consultants led the way in asking the wireless department to put on a pair of American shoes as a pilot project. After observation, the experiment was extended to all product lines in Huawei. However, after "cutting their feet down to size to fit the shoes," the wireless department soon felt enormous pain.

It had earlier discovered that there was a tremendous potential market in "instant message," and that it might even become mainstream business in Internet telecom, following the end of traditional telecom. The department worked night and day to complete a design. After the review process, however, the idea met with refusal. In contrast, Tencent was able to grab hold of this historic opportunity, and that company became a legend in modern times. After that, in the process of "cutting the feet to fit the shoes" on a widespread basis in the company, the company also came up against a global IT freeze. A large number of core R&D staff, unable to bear this "cutting the feet to fit the shoes," then opted to leave and look elsewhere for work.

This western-style experiment was unprecedented in Chinese business history. In the course of it, the ones hurting were not just several tens of thousands of Huawei employees, but also Ren himself.

From 1996 to 1998, Huawei revenues rose ferociously from RMB 2 billion to RMB 8.9 billion. After 1998, they then "exploded" in 1999 and 2000. Customers dispatched orders to Huawei in a frenzy, and business was extremely hot. As per habit, sales personnel agreed to try to satisfy the needs of customers. Unlike before, however, the R&D department was unable to push out product or the corresponding product solutions as fast. A large number of potential orders in hand therefore had to be relinquished to competitors.

Ren's statement that "We will continue to keep in mind the dream of creating our own Chinese model and Huawei's own model" had been very moving to people. His courage in attempting change had amazed everyone. Meanwhile, IBM had promised to achieve change that cut to the very bone within five years. In those five years, however, high-tech was upgrading to new systems on an average of once every 49 days. Competition was intense. Who could have known that within five years the information age would go through earth-shattering change itself? What's more, all the various changes being undertaken at Huawei were just being applied to the "dragon head" of R&D. Behind it came the "dragon body," with all the various departments. These too needed "major operations." For Ren, the question of whether the patient would die on the operating table was a very real one.

In July 1998, Ren had said in his speech on not being a short-blooming hero, "The hardest thing to change is your own life. To have others try to change your life is much easier than for you to try to change it yourself." His intention was to remind Huawei people that they had to prepare in advance, prior to even trying to achieve total westernization of processes. Facts were again to prove that it is impossible to transform people's way of thinking, however. Or at least the degree of difficulty far surpassed what Ren had imagined. It was at this point that he began to be aware that he had already set sail on a journey and could no longer "sail backwards." However, sailing forward was being blocked by the enormous force of culture. Meanwhile, of critical importance was the fact that a shockingly intense global Internet winter arrived just as he was trying to cut Huawei's feet to match western shoes. This was undeniably a major test of his determination to take a "one-third share in all under heaven," as well as a test of his courage in trying to change management.

CHAPTER EIGHT

HUAWEI'S WINTER

Huawei's crisis, its contraction, and then its bankruptcy, are inevitable. Everybody must prepare for it. This is my unshakable belief since it is in accord with the laws of history.

—REN ZHENGFEI

Objectively speaking, Huawei's success in the early period can be attributed to three main factors. First, Huawei seized a historic opportunity as China spent hundreds of millions every year to put in fixed landlines and build up the country's telecom infrastructure. Second, it had its own self-generated core technology. Third, it had the charismatic power of Ren himself.

In addition to the usual attributes, self-confidence, self-discipline, and so on, Ren had other features that became apparent in the sharp-edged turbulence of intense competition. He maintained the highly acute instincts of a "commercial politician" in terms of foresight and a sense of positioning. In every decision, he seems to have been able to make things happen that looked impossible.

In this regard, he once said in an internal meeting in the company, "In the course of these past 10 years, we have fortunately been able to resolve every foreseeable problem with measures that were in step with the changing times. The rhythm was in sync, which is how we could achieve success and become what we are today."

For example, in 1993, Zheng Baoyong and Li Yinan were in charge of the R&D for Huawei's C&C08 switches. They believed that the market for the 2000 line of switches was pretty much finished, but Ren persisted in believing otherwise. He had the R&D department continue to develop a 10,000 line and then a 100,000 line. The market could not get enough of these, and ultimately they became mainstream models in China.

As another example, in 1994, when the rest of the industry's manufacturers were still steeped in the red-hot era of control-process switches, Ren led the way in asking Zheng Baoyong to establish R&D facilities in Beijing and Shanghai to research digital telecom technology. This move was later regarded as highly prescient. In 1999, the switches market began to decline as digital telecom suddenly took off, and it went on to become mainstream business for operators. Huawei was catapulted into an era of high-speed growth.

Long-term strategy was consistently aiming the company in the primary direction that the market was taking, while short-term decisions had, to date, mostly been correct as well.

At the end of 1996, Ren instructed the procurement office to purchase RMB 1.5 billion worth of original components. At the time, many senior managers reminded him that these components were extremely time sensitive. As soon as products moved into new-generation technology, or policies changed, Huawei would be burdened with an overstocked inventory, and they feared this would increase the company's risk of going bankrupt. Ren nevertheless came out with a statement that few could believe: "You just watch. Within three years, our sales will quadruple." As it turned out, in 1997, Huawei realized sales of RMB 4.1 billion. Moving into 1998, the sales volume seemed unstoppable, and in the third quarter, sales for that year already exceeded RMB 8 billion.

The state of the industry at that time was such that you could practically sell as much as you could produce. This enormous market demand also led to soaring prices for original components. Many competitors had to stop production since they could not get enough raw materials. Other competitors had to buy components at a high price and then tried to change their contracts with buyers, which led to disgruntled customers. Huawei, with plenty of ammunition in its hold, could occupy large swathes of the market due to its cost advantage, but it also took in many orders that other manufacturers could not supply on time.

Ren's story of hoarding RMB 1.5 billion worth of goods became a legend within the company. It was also broadly recognized as being a turning point for the company as it moved out of the strategy of "using the countryside to encircle the cities."

An Age of Rocketlike Growth

After these successes in both fixed-line and digital telecom, Ren ambitiously began to focus his attention on the mobile telecom arena.

In 1997, in a country with a population of 1.26 billion, the penetration rate of mobile telephones was not even 0.5 percent. In 1998, the projection was that this would rise to around 1 percent. Meanwhile, during that period, according to data of the International Telecommunication Union (ITU), mobile customers around the world totaled 281 million in 1998, 400 million in 1998, and 1 billion in 2000. These numbers provided an extremely magnificent prospect, whether you were thinking of China's market alone or that of the entire world.

To seize this huge growth opportunity, Ren first organized an R&D team with "wolflike" attributes. He asked for mandatory breakthroughs in GSM technology, using "pressure principles." He instructed the human resources department to hire telecom talent from universities as well as the public at large, in order to have sufficient personnel on hand to meet peak demand in the mobile telecom initiative.

In the second half of 1997, Huawei hired tremendous numbers of people from institutes of higher education including Beijing's Post and Telecom University, Nanjing's Post and Telecom University, and Xi'an Electronics Technology University. Some telecom and computer departments were simply hired en masse. It also paid high salaries to attract new hires that were skilled in telecom and marketing in such cities as Beijing, Shanghai, and Xi'an. In the space of two years, the human resources department at Huawei hired close to 10,000 people. In 1997, Huawei had more than 4,800 employees; by the end of 1999, this had gone up to 20,000.

Meanwhile, as Huawei broke into GSM technology and had its technology certified at the national level, the company began to see a ray of

light in international markets with wireless products. As Ren accelerated the promulgation of *Huawei Basic Law*, and as he brought in new
management systems with IBM as the representative of European and
American models, he set the goal of RMB 16 billion in sales for 1999.
This was based on Huawei's three successive years of doubled growth
figures. Meanwhile, he dispatched his most experienced marketing personnel on marketing trips to explore overseas markets. They went one
by one to countries in Asia, Africa, and Latin America to initiate "large-
scale offensives."

Problems began to appear just as all the pieces were in place to implement his strategic vision.

The IT Bubble Bursts

From the mid-1980s to the late 1990s, the torch of the IT industry was
ignited and lit up the entire world under the impetus of what was seen
as the "new economic theory" and "virtual wealth" emanating from the
United States. Anything related to "networks" seemed unbeatable. Many
listed companies relied on the idea of a "dot com" to make millions
with the greatest of ease. Any extreme comes back on itself, however. As
finances and talent flowed into the IT industry on a global scale, traditional industries lacked both funding and talent, and their growth slowed
in relative terms. As demand for "IT-ized" products sharply declined, this
led directly to a glut in human talent, products, and markets. Ultimately,
the extremely high pricing structure of the IT industry caved in under
fierce competitive pressure.

In the first quarter of 2000, Nasdaq's overall P/E ratio swiftly declined.
After floating high at 300 percent, it plummeted as the weather changed.

On March 10, 2000, Nasdaq declined by 50 percent. On April 10,
it fell by more than 80 percent. One week later, "network shares" were
approaching collapse. In Silicon Valley, thousands of IT companies were
forced to declare bankruptcy overnight. Meanwhile, the dreams of wealth
that millions of shareholders around the world had entertained were also
turned to dust in a moment. Given the "irrational exuberance" with which

people had madly invested in the IT arena, the torch that had burned so brightly was suddenly extinguished.

A close supply and demand relationship existed among all parts of the IT industry including operators and telecom equipment manufacturers. Millions of threads bound them together. After the IT industry went belly up, this led to a string of bankruptcies all down the line, which made economic conditions even worse.

One example was Global Telecom, an American company that was established in 1997. Through a series of mergers and acquisitions, and by taking on enormous debt in the period of booming IT, Global Telecom's optical network covered three continents, 27 countries, and more than 200 cities. It became a leader in the global IT buildup industry. In May 1999, at its height, its shares were worth USD 47 billion on the market. After the IT bubble burst, its market value swiftly declined to USD 750 million. In the end, the company was unable to weather the crisis and declared bankruptcy due to the cost of carrying its debt.

More than 80 percent of the optical products for Global Telecom had been supplied by Lucent, Nortel, and Marconi. After Global Telecom went bankrupt, this led to an enormous amount of unrecoverable debt. As creditors, these three global giants in telecom became mired in losses as well. According to publicly announced figures in 2001, Lucent lost USD 16.2 billion and Nortel lost USD 25.7 billion. The shares of Makeny suffered a tragic fall from a high of 12.5 pounds to 6 shillings. In addition, other European giants in the telecom sphere were heavily wounded, and losses hit each one to varying degrees.

During a trip to the United States, Ren learned from friends in the IT industry that global losses from the bursting of the IT bubble resulted in eliminating some USD 20 trillion in wealth overall. Of this sum, the United States had lost USD 9 trillion, USD 3 trillion of which was specifically in the IT industry. The tragedy being played out on Nasdaq right now, they said, was also just the tip of the iceberg. It would take Silicon Valley at least 10 years to recover.

Prior to this, Ren had frequently expressed the opinion in internal meetings that the abnormal growth of the IT industry violated the laws of value. Sooner or later, it would lead to disaster. He had not, however, anticipated that the disaster would come so swiftly and so tragically. Lucent and Nortel were particularly hard hit, and no European company escaped being rolled up in the process. As he considered this, he had a premonition of what was to come.

The Fall of Internet Shares and the Ripple Effect on Huawei
Given the rate of technological changes in the telecom field, as well as overinvestment and intensified competition, the telecom market in Europe and the United States was already trending toward saturation. In contrast, the market in China, with its population of 1.2 billion, still had enormous potential. Mobile telecom was an example. Both international and domestic forecasts of market size[1] said that the number of mobile users in China would continue to grow at a rate of 20 percent per year between 1999 and 2003. What's more, as China entered the World Trade Organization (WTO), opening up the telecom market was going to be the main trend with European and American giants trying to emerge from their trauma by carving up the China pie. Foreign companies were moving both R&D centers and production facilities to China, using China's cheap labor to lower their costs and raise competitiveness. Huawei's innate advantage in lower costs would therefore be threatened in the new round as telecom companies locked horns with one another.

Although China's telecom market looked prosperous on the surface, this was in fact the result of the rise in Internet stocks over the past few years. Now, with China impacted by the dramatic fall in those stocks, there was inevitably going to be a sharp pullback in investment in China. The buildup of Chinese telecom was going to take a hit as the global economy worsened, which would lead to a period of contraction.

"This wintertime is going to be as surprisingly cold for Internet suppliers as the hot times were unbelievably hot."

After completing his trip to the United States, Ren put his thoughts on this into what became a famous article called "Huawei's Winter."

His premonition was soon to be proved correct.

---

[1]The international forecast came from the International Telecom Union and the domestic Chinese forecast from China's Ministry of Information Industries.

After 2000, as Internet stocks declined precipitously, two Chinese companies, telecom operators that had been listed in Hong Kong and New York, also began to see a fall in their share prices. Those were China Mobile and China Unicom. Their shortage of funds directly impacted their willingness to invest in equipment. In addition, the fixed-line telephone market was now becoming saturated due to more than 10 years of abnormally high growth. The network buildup in various provinces moved toward "improvements" rather than ongoing investment. What's more, China's Ministry of Information Industries now decided to divide China's telecom market into north and south. Once this rumor began to spread, but had not yet been officially confirmed, a number of uncertainties made both sides sit quietly on the sidelines until everything became clearer. On the south side was China Telecom, and on the north side, China Netcom.[2] Some projects were already in the pipeline—they now tried to delay things and reduce the size of investments. Those projects not yet in the pipeline simply cut out infrastructure building altogether. In 2000, therefore, the three main operators in the telecom market in China were in a state of semisuspended motion as far as any investment into infrastructure was concerned.

This sudden change in market conditions led to a steep reduction in income. It put the costs of Huawei's "grand hiring of 10,000 people" into sharp relief. The cruelties of the reality of the situation meant that Ren now had to find other profit-making areas outside the shrinking telecom market.

Start-ups Within the Company

Since its founding, Huawei has set its feet firmly in the turf of telecom markets. What's more, it has done well by a direct sales method. Huawei appointed a highly competent person named Yuan Xi to head an enterprise network department. This aimed at the enterprise network market in various fields in China, including the military, oil fields, electric

---

[2]China's telecom industry was again restructured in 2008. There are now three main companies, all state-run: China Telecom, China Unicom, and China Mobile. China Unicom acquired China Netcom in October 2008.

power, transport, coal, government, and education. Since Huawei had never cultivated alternative distribution channels, however, and since it focused on a direct-sales method, the company never did that well in the enterprise network market.

After the IT bubble burst, global telecom markets began a period of contraction, but the impact on enterprise network markets was less notable. IBM and Cisco were never hurt as badly as the giants in telecom industries in Europe and America. On the contrary, their operating income went against the tide and actually increased. The main reason was that in addition to being in telecom markets they were in enterprise network markets. Moreover, they had constructed a dense network of distribution channels worldwide, which provided powerful support for ongoing stable growth. For example, in 2000, Cisco realized USD 500 million in sales via distributor channels inside China to the enterprise network market. In contrast, Huawei's share of that market in China was minimal.

Going into the year 2000, Huawei was already taking the initial steps to form an embryonic kind of IPD and ISC in the company, as its core business procedures. This was under the unstinting efforts of IBM consultants and employees as well as Ren's "iron wrist," making it happen. In the process, some cadres in the middle ranks of management became very wealthy people. Most of these were "old employees" who dated from the company's founding. They had fought on the battle lines for years, and were extremely hard-working. Ren now launched an initiative that was called "start-ups within the company," which allowed these old-timers to go out on their own. The idea was to enable these people to start serving as agents for Huawei products. They had both "great merit" and tremendous experience. They could realize their dream of being "bosses" themselves, while at the same time making up for Huawei's lack of distribution channels and the disparity it had with Cisco in this regard. Together, they and Huawei could build up the market in enterprise networks. Without doubt, this was an excellent plan as a way to help get Huawei through the IT winter. This plan, highly "humanistic" in nature, had been gestating in Ren's mind for two years. Over that time, he had begun to lay the foundation for it and to test it in various ways.

On December 2, 1998, Huawei's journal *Better Management* published an article entitled "Notice on Various Issues to Do with Launching an Employee-Run Internal Start-up Initiative." This stated that employees who had worked a full three years at the company could sign contracts to "outsource" various parts of Huawei's procedures, including printing, express delivery, laundry, cafeterias, internal coffee and gift shops, packaging that had to do with Huawei equipment, and so on. To address the concerns of family members and encourage employees to be proactive in such start-ups, Ren invited many families of employees to direct discussions that explained the advantages of these internal start-ups.

With the direct approval of Ren, a group of older employees quickly signed contracts for Huawei's printing, packaging, and shops. They became the first batch of "start-up entrepreneurs." Their business developed quickly owing to Huawei's own strong growth, and they also amassed considerable expertise in running a company. After more than one year of pilot projects, Ren discovered that the results of these internal start-ups were quite respectable. As the icy winds of winter approached, he therefore accelerated policies in this direction.

On August 16, 2000, Huawei's journal *Better Management* published another article, called "Management Rules Regarding Internal Start-ups." Any employee who had completed a full two years in the company could now apply to start a company and become an agent for Huawei. To support employees with their start-ups, the company would provide, at no cost, equipment up to 70 percent of the value of the employee's internal shares. If the employee's company failed within one-half year, she or he could return to the company and have a new position.

As soon as this policy was announced, nearly 1,000 employees began their own start-ups. They had the strong backing of procedures that ensured they would have enough "to eat" and did not have to worry about a berth in the future. Among them were Li Yinan, head of the R&D department, and Nie Guoliang, vice president of Huawei. Full of excitement and ambition, each stepped onto the path of being an independent entrepreneur. However, a move that had been initiated with such good intentions and such hopes turned out to become an unprecedented disaster for Huawei. It was one that plagued and saddened Ren for a long time to come.

A Tragedy that Came Out of the Blue

On January 8, 2001, Ren was overseas on business. He suddenly received a phone call from the head of the finance department, informing him that Ren's mother had been in a car accident in Kunming. Ren set off immediately. He flew from Iran to Brazil, where he had to wait six hours for a connection. From Brazil he went to Rangoon where he was delayed another two hours due to heavy rains, so he missed the connection on to Beijing. He reached Kunming only on the evening of the next day, heart full of anguish.

By that time, his mother was on life support, with her heart and her breathing controlled by medical equipment. She saw him one last time and then suddenly passed away.

Ren's father had died in 1995. At that time, Huawei had only begun to be more "standardized" as a business. Worried about Ren's state of mind, the family kept the news from him as his father's condition worsened. Only when his father was put in the hospital did they notify him. Ren sped to Kunming where his father was already in a coma. Only now that father and son were at this final parting stage did Ren suddenly remember how his father had never asked him for a single thing, and Ren in turn had never even bought his father a single piece of clothing. At the funeral, stricken with grief, Ren took off his jacket and spread it over the old man as he lay there, a man who had been so very frugal throughout his life.

In the days to follow, Ren constantly reminded employees in the company to make up for lost time and be filial to their parents. He particularly encouraged the male employees, more "hard-hearted" in their manner, to show respect for their parents in any way possible. After his mother died, at the end of every annual awards ceremony, he would admonish all present, who would soon be going home for the holidays, "not to forget to wash the feet of your mother and father." He himself hoped to be able to accompany his mother in her older days, once the company was on sound footing and he could retire. He had no idea that tragedy would hit him crosswise.

After arranging his mother's funeral and putting his affairs in order, Ren soon was hit by diabetes, high blood pressure, and other illnesses brought on in part by mental trauma. He lost energy, grew thin, and suddenly looked haggard and much older than his 57 years. The chairman of the board, Sun Yafang, and several members of senior management met to discuss how to make him take a break to restore his health. The following message soon appeared on the company's internal network:

> *On the 8th of January, 2001, Ren-zong was on a business trip to Iran when he suddenly learned that his mother had been in a car accident in Kunming. She passed away on the January 10th. In the days since this unexpected calamity, Ren-zong's family, friends, and colleagues have been extremely concerned about him and have helped him and supported him. On his behalf we express his sincere appreciation. More recently, some customers, suppliers, and partners have also learned about this and constantly tried to express their condolences. Ren-zong is a rational man, but he is hurting inside and this may continue for three to five years. The more he gets phone calls and expressions of sympathy, the more he continues to be traumatized. Because of this, we ask employees inside the company not to bother him, and we ask people outside the company to refrain from bothering him as well, on principle. On his behalf, we express appreciation for reducing the psychological pressure on him right now.*

For the next half month, Ren was immersed in profound grief. He felt he had not fulfilled his responsibilities in caring for his parents. He wept as he wrote down his thoughts about how he had not taken care of them when he couldn't really do so, but he also had not taken care of them once he was able to do so. He regretted that now. He regretted not taking the time to be with them.

This memorial, which was quite poetic, was later published in the internal *Huawei* magazine and elicited a powerful response from everyone in the company as well as others who saw it. It ended with an oath to carry on. Those who were gone were now simply gone. There was nothing the living could do but pull together and move forward. Many people cried when they read the stanzas and realized it was written from the heart of a man who was a high-tech kind of hero but who also was deeply sad that he had not fulfilled his duties in the more spiritual realm.

On February 28, 2001, Ren emerged from his profound grief to call an emergency annual meeting of management personnel at the level of divisions and higher. At this meeting, he delivered a report aimed at Huawei's current situation, called the "Huawei Ten Major Management Issues." In the mind of the industry, it also became known as "Huawei's Winter."

"For 10 years," he wrote, "I have lived daily with the idea of crisis. I saw the successes but did not register them, nor did I feel any sense of glory or pride in them. Instead, what I felt was a sense of crisis, which may be why we have survived these 10 years. Huawei as a company has always been crying 'wolf.' We cried too much, perhaps, and nobody believed us. Well, the wolf is on his way."

It took Ren three hours to deliver this report. The entire document, from start to finish, revolved around one main topic, namely, crisis. However, employees had long since become accustomed to having Ren drill them in the correct way of thinking. From the day it was founded, virtually every article, speech, forum, and report had reiterated the potential for crisis. On April 3 and July 20, 1999, as the new employees were introduced to the company, Ren repeated this theme: "Huawei could collapse as a company at any time. What exactly is crisis? Crisis happens when we still don't know where it is, when we have no sense of it. If we knew what and where it is, Huawei would not face it at all since we could then take measures to deal with it."

This time, however, these warning words came at a time when the global telecom market was on the verge of dramatic decline. The hammer was hitting the bell at a particularly resounding time. Ren himself, however, had no idea that, after 12 years of high-speed growth, "crisis" would bare its fangs at the company in ways he was most unwilling to see.

The Mistake of Xiaolingtong[3]

The first time Huawei began developing end-user products was in 1994. Once the C&C08 switches came out, Huawei then began to develop all kinds of specialty phones.

In 1995, after the "mass resignation" of the marketing department, the marketing department of Huawei's end-user business department began promoting a different kind of "talking machine." Since it was not high quality, however, and also not priced competitively, its market never rose to expectations. In 1996, the end-user department developed an integrated services digital network (ISDN) series of end-user products, to go along with the C&C08 process-control switches as a bundled product in the marketing. Although this time the R&D orientation was correct, the ISDN technology was quickly superseded by an asymmetric digital subscriber line (ADSL), and eventually the demise of ISDN was marked by considerable losses.

These two failures taught Huawei a lesson: In the field of control-process switches, the competitors were from abroad, so Huawei had an edge due to its cost advantage. In end-user products, however, the tech content was fairly low so it was hard for the company to be competitive. Even if the company produced at a scale that made it first in the business, profits would still be very slim. As a result, Huawei decided not to get involved in end-user products.

In the mid-1990s, wireless telecom manufacturers, most notably Nokia, had massive success in the GSM arena. After that, based on their brand advantage, these companies swiftly rolled up the global market in cell phones. A number of other companies then swiftly plunged into the cell phone market, including Motorola, Lucent, Ericsson, and Zhongxing [ZTL] among other companies within China. Faced with its recent failures, Huawei instead failed to pay enough attention to this shift in the industry.

---

[3]Halfway between mobile and fixed, Xiaolingtong is a limited-mobility service based on personal access system (PAS) and personal handy (phone) system (PHS) technology. It consists of a wireless local loop that provides access to the fixed-line network.

Around the mid-autumn festival in 1998, the head of the Hangzhou office of Huawei, Dr. Shao Shuwen, learned that the Yuhang telecom bureau of Zhejiang province was promoting the xiaolingtong-type wireless phones on the market, and he immediately passed this on to headquarters. However, when it came to setting up a project to pursue this, there were differences of opinion on the future prospects for xiaolingtong. In the end, given Ren's decision, Huawei chose to put its main forces into 3G and not to pursue the strategy of developing Xiaolingtong.

In 2000, once China Mobile was spun off from China Telecom, its year-end revenues swiftly surpassed China Telecom's 15 percent. The younger brother overtook the elder brother, which put tremendous pressure on China Telecom simply to stay alive. As the market for fixed-line telephones shrank by the day, the market for mobile telephones predominated as forces rushed into the market. The core issue for China Telecom now became how to open up new space for itself in the wireless business, before its 3G license was issued.

Faced with a great deal of controversy, and after much study and discussion, China's Ministry of Information Industries issued a document with the following regulation: China Telecom was not allowed to engage in mobile business, including GSM and CDMA, but wireless products could be regarded as an "extension of and a supplement to fixed-line telephone services."

This additional opportunity was extremely welcome to China Telecom. Since Xiaolingtong had been successfully used commercially in the Zhejiang Yuhang Telecom Bureau, this gave China Telecom particular confidence. Prior to the time the GSM license was issued, Xiaolingtong could satisfy the needs of the common person in China. Ultimately, as telecom policies were determined and made public, the whole issue of licenses for various standards was to erupt like a volcano.

In 2001, without any particular promotion, China Telecom fairly easily gained 12 million customers for Xiaolingtong. On through 2002, 2003, and 2004, the Xiaolingtong network capacity reached between 6 and 7 million as the business grew ferociously. During this period, China Telecom and China Netcom invested the huge amount of more than RMB 90 billion in Xiaolingtong equipment. Since Huawei did not have a seat in this arena, the Xiaolingtong business was split among three companies, UTStarcom, with 60 percent; Zhongxing Telecom, with 30 percent; and Qingdao Lucent, with 10 percent. Meanwhile, in the midst of its own winter, Huawei simply brushed shoulders with these entities and passed on.

CDMA

In 1996, Huawei had two choices as it decided to take its forces into the field of mobile telecom. These were the two mainstream international technologies: CDMA and GSM. CDMA was the American system. With good connectivity and slight radial variation, it had considerable technological content. GSM was the European system. Also known as Global Connect, this technology was slightly inferior but the R&D involved was also simpler.

Huawei was financially incapable of having two systems running in parallel, so Ren made the decision and chose GSM. He then assembled a team of experts to do the research and development. Within two years, they had broken through the technology and had the results appraised and approved at the national level.

In February 2000, as the date approached for China's formal entry into the WTO, China and Qualcomm signed a "framework agreement" for investing in building CDMA.

Once this news became known, Huawei, which did not have CDMA, urgently pulled together several hundred developers for an all-out effort to "break through" or reverse-engineer, the technology. During the spring holiday period of 2000, they shut themselves into the offices on Shiyan Lake in Shenzhen and worked day and night to crack the code. Achieving a complete breakthrough was difficult due to the CDMA technology itself. On May 21, 2001, therefore, Huawei decided to join forces with Motorola when China's Unicom CDMA issued a call for bids, with as much as RMB 100 million in business at stake. Since many of the functional standards did not meet the requirements of the Unicom network, however, Huawei was only able to get some orders, for mobile switches. After the lessons of this defeat, Huawei renewed its efforts and fought for another year to crack the technology. Not only did it then break through the full set of CDMA technology, but also it put out a higher version of CDMA 2000. Various provinces and municipalities in China tested the technology and confirmed it as being at the cutting edge of international standards. Since Huawei had not won the first tender, however, it was unlikely that Unicom was going to choose it on the second tender. The company was not going to dismantle everything it had bought in the first tender and switch it over. Therefore, as Unicom CDMA finished its second tender in November 2002, for business in excess of RMB 20 million, Huawei lost out yet again.

Going from "Internal Start-ups" to "Internal Injuries"

Ren had mobilized the internal start-up campaign in part to create an enterprise network market that had Huawei as its central axis. But more importantly, he wanted to provide older employees with room to grow, after they had dedicated such effort on behalf of Huawei. The timing was not good, however, due to the global "winter" that was afflicting the IT industry as a whole. Moreover, in October 2001, an American company called Emerson purchased Huawei's electric power business, and after this, rumors began circulating that Huawei was going to be split into separate components, some of which would be listed on the market. This made those who had intended to run "internal start-ups" even more cautious in how they proceeded.

Most of the employees who had left their corporate positions and started their own entities were in fact extremely capable. They were quite able to realize their own value as they founded companies. Li Yinan was a particularly good example. Once he went north to start his own company, a number of other backbone cadres started up companies in R&D as well as marketing. After that, between December 2000 and the end of 2002, a large group of internal start-up companies sprang up, all serving as agents for Huawei products.

Market competition was extremely intense owing to the challenges of the IT winter. This adverse environment forced some of the smaller companies to adopt all kinds of ways and means to survive. These included taking advantage of Huawei's platform resources in order to share in

Huawei's markets—this sort of thing was simply hard to avoid. It had a negative impact on the work ethic of people still employed at Huawei, however. It also shook their faith in the company.

As the winter intensified, sales volume at Huawei dropped off precipitously. This was precisely when the movement that Ren had initiated, "cutting the feet to fit the shoes," was now beginning to bite into the flesh. In this most difficult and painful period, this most critical stage, quite a few employees not only left but also took the company's intellectual property with them. They took the original code, design blueprints, and various other forms of technical secrets. They either made products themselves or sold these things to competitors. This kind of zero-cost, zero-investment piracy of Huawei's products then went head-on into competition with Huawei in the market. It almost drowned Huawei in the process.

On May 10, 2006, after purchasing a Hong Kong company called the Harbour Network, Ren met with some of his former senior managers, including Li Yinan and Huang Yaoxu, and described the situation. "From 2001 to 2002, Huawei was being hit both inside and outside and was on the brink of collapse. At that time, those who remained in the company were looked on as somehow being abnormal. A very unsavory trend permeated the company. Everyone began to say that 'the early-period capital was dirty.' They said that we had been funded by dirty money. It then almost became a point of honor to steal company technology and corporate secrets, particularly when venture capital was providing such incentives to these people. The turbulence at that time almost toppled the company."

CHAPTER NINE

DEALING WITH TURBULENCE

*Huawei does not say that it is going to enter the ranks of the
World 500 within three generations of people. Yes, we say, "three
generations," but we mean the three times Huawei has gone
down and then reemerged, not three generations of leadership.*

—REN ZHENGFEI

As he explored how to turn Huawei into an international company, Ren
Zhengfei put a tremendous amount of energy into researching Japanese
companies, not just American companies.

The 1960s and 1970s were the age of electronics. Value added came
mainly from manufacturing, and Sony, Matsushita, Fukuda, and Hitachi
became the most famous among tens of thousands of Japanese companies
who made up the core arena of global manufacturing. Japan's economy
soared on this basis. At the time, the country was the very epicenter of
profitability in the world.

In the 1980s, the United States adopted more open policies. Aided
by a booming stock market, this brought in outstanding talent from
around the world. Technology, funds, and personnel led to the phe-
nomenal growth of information technologies to the extent that the
United States soon held more than 60 percent of the IT market. In part
due to this advantage, the country became the number one economy
in the world.

Meanwhile Japan was steeped in its earlier massive successes; and
this, in addition to a uniform ethnic makeup, an already prosperous

way of life, and the lack of competition, held back innovation and any real determination to change. As information technologies entered into industrial systems, the value added in the business quickly shifted toward researching and developing core technologies and toward links in the sales process. Japan, with its competitive advantage in manufacturing, was badly hurt by this shift. It quickly became marginalized as it slipped away from the center of profitability. Starting in the early 1990s, Japan entered into a prolonged period—10 years—of low growth, zero growth, or negative growth.

Japanese companies had been massively successful, but they could not maintain long-term growth by relying on just one kind of structure. One had to wonder if Huawei's previous successes could be relied upon as a firm model for the future. Meanwhile Japan's companies had been undergoing a very long winter and yet hanging on. As they did so, Ren wondered how they endured, what problems they came up against, and what lessons they could teach others.

After going through a fairly smooth development period, how was Huawei to deal with the coming challenges? To answer this question, Ren's thoughts first went to Japan.

Entrepreneurial Spirit and "Springtime in the North"

Japan was still in the midst of spring in April 2001, when Ren caught a flight to the country to find the answer to his question about how to survive a brutal winter. Cherry blossoms were blooming and sunny scenes lit up the countryside.

In a small *ryokan* in a remote village where he was staying, Ren observed a group of retired older people enjoying a gathering, as they chanted a folksong about pulling in the fishing nets. He noticed how optimistic they were, how full of enthusiasm, how unworried. He could not help but be reminded of his father and mother, and sadness came over him. "Old people in China are burdened down with troubles. Their lives are exhausting. People in our previous generation lived under such a heavy state of mind that they died, never having been able to lighten up."

"Japanese friends translated the piece I wrote about my parents into Japanese and also English, and asked employees to read it. This mistakenly assumed that I was a filial son. In fact, I was the opposite—I did not fulfill my responsibility to my parents or I would not be as troubled and hurt by their deaths as I am. I put all of my efforts into work and neglected my parents—in fact I hardly deserve to be called a son at all."

When the song about pulling in the nets was over, prolonged applause interrupted these thoughts of Ren about his parents. But they then began singing a famous song about the road to Hokkaido, called "Springtime in the North." The simple unadorned lyrics and the soulful melody now deeply touched Ren's memories about the hardships undergone in starting the company. Under his breath, he too hummed this much beloved tune as tears rolled down his cheeks.

"When a young person 'turns his back to the well and leaves his homeland,' goes far from his family to struggle to make a living, only his mother thinks of him at every moment. She sends him a cotton-padded jacket not knowing that springtime has already arrived in the big city."

"Any of us who had the opportunity to go to college went because our fathers and brothers sacrificed themselves on our behalf. Their hard work built up our chances, bit by bit. They worked to provide for the wandering son, so far away and alone. They themselves had no education. They used their own strong backbones to do the work. For us, they built up the first step that we might take that step and lift ourselves up further. We absolutely must not forget them. We must not think lightly of them, must not forget to pay them back."

"Each person's success in fact comes from the selfless contributions of his family. The original driving force of our lives, our work, our endeavors, comes from our mothers, who made us warm winter clothing, from our silent fathers and elder brothers, and from that young lady who may once have loved you and from whom you are now separated..."

This song, pregnant with significance of all kinds, resonated within Ren. It brought back old memories, and it made him think of just how patient and resilient the Japanese people were, how they had an optimistic spirit. He could not help but think that the song was an expression of the very fighting spirit of the Japanese people. Based on that spirit, they had shaken the world with their electronics in an earlier age. Moreover, their uncomplaining vigor had been put to work for Japan's glory.

"Although Japan has currently run up against problems, the forbearance, optimism, diligence, and fighting spirit of the Japanese people have not changed. Their beliefs have not changed. Their enthusiasm for life and for work has not changed. When Japan's economy takes off again, this foundation will allow it to reach for the stars."

Solutions for How to Deal with Winter

As he walked through Japan's streets, Ren's impressions were similar to those of 10 years earlier: quiet, congenial people who were clean, prosperous, and comfortable. Everyone seemed generous and polite. People moved quickly, whether they were the "working tribe" on busy streets or the waitress in a noodle shop. In the midst of equanimity, optimism, and respect, they seemed to treasure their work and the chance to serve others. By this time, Japan had already been in the longest decline of any time since the end of the war, 10 long years of winter. For close to eight years, most companies had not raised wages, but social order seemed more intact than in northern Europe.

Ren saw all this and his thoughts turned to Huawei. "Huawei went through 10 years of high-speed growth. Can we keep that up or will we face slow growth, and even protracted slow growth? What problems do we need to deal with in our corporate structure and in management?"

"Employees were promoted briskly in 'peaceful' times, but can they cope with the severity of winter? If we meet up with a succession of winters, will they stay quiet? Do they have the ability to deal with problems, overcome them, and wait for springtime?"

While visiting Matsushita, Ren discovered that every wall, in bathrooms, offices, and corridors, was hung with the same photo and inscription—a large ship about to crash into an iceberg with the caption, "Only you can save this ship." Obviously, the "sense of crisis" in this company was high. All that he saw touched a chord, which made him wonder: How powerful is Huawei's own awareness about impending icebergs? Can we transmit our concerns to the grassroots level? Can we mobilize each and every person?

"Huawei went through too long a period of calm, and we promoted too many managers during that period. This may constitute a disaster for us in the future. The Titanic too left on its voyage with a great shout of glory."

While in Japan, Ren met with the famous Japanese scholar of management studies and consultant on national affairs, Takeuchi Rinki. Takeuchi said quite frankly, "Any person, enterprise, or even any country may well have a period of decline after its ascent reaches a certain point. For 10 years, we have consistently hired too many people, put in too much equipment, loaded on too much debt. This burden now weighs on Japanese companies as a major lesson. It reveals an objective law: anyone who simply extends what he relied on for success in the past will inevitably face death."

The trajectory of Japan's evolution from glory to decline was apparent, and the parallels were obvious. Ren was later to declare the following:

Huawei grew up when the world's information industries were growing at the fastest period in history. What's more, China went from having a totally backward network to having a modern, world-class network at a very fast pace. Huawei was just like a leaf on this huge wave. It very fortunately fell onto a huge ship that was navigating the flow, and it stayed hidden there, riding up and down as the ship handled the waves. It did not experience the froth of those waves itself, however. It did not feel the force of the flood or the smash of waves. Because of this, Huawei's success was more about luck than it was due to talent or native ability. The moment the rate of growth in the company slows down, like Japan, Huawei is going to face crises. Like Japan, we have hired too many people, bought too much equipment, and taken on too much debt.

He went on to say the following:

It has been hard for all Asian companies to go "international," but particularly hard for China. China was closed for decades, and it has been a mere 20 years since reform and the opening up. This is not enough to support the country's internationalization. First, there is the problem of aligning management with international practices. Quite apart from that, the language barrier is substantial. Huawei has large numbers of foreign nationals among its workers, who cannot read Chinese documents, and it has numbers of employees within China whose level of English is not good enough to pass muster. These things alone show how hard it will be to internationalize the company. Meanwhile, Huawei also has a host of management problems. It would take days to even start listing these. If we do not take urgent action in overcoming these problems, and continue to press on with reforms, then Huawei's "glory" will be a flash in the pan. As times change, the company will be put on the defensive and ultimately will be forced to cease to exist.

Concluding this trip to Japan, Ren wrote an article called "Springtime in the North," in which he sought to encourage employees and in which he described what he felt were some of the solutions to Huawei's winter. "What is true success? It is being able to go through nine deaths and still be alive, like the Japanese. That is true success. Huawei does not yet have success. What it has is growth."

"Founding a company, and keeping it going, are hard. Winter will eventually be over, however. As we ride through it, we must anticipate spring. We must use all we can to prepare for when the snows melt and the brooks start running again. Anticipating means strengthening ourselves internally and it means cultivating the right spirit. Springtime will eventually arrive."

USD 750 Million to Purchase Huawei's Electric Power Subsidiary

In providing his employees with a "warm padded coat" for their way of thinking and their spirit, Ren was also beginning to put major effort into making the coat that would prepare Huawei for dealing with winter.

Huawei Electric was the company's sole subsidiary. In 1996, as Huawei broke into the China telecom market and grew tremendously, Ren spun off the production line for batteries and incorporated it into Huawei Electric, with independent operations. After several years, based on the backing of its mother entity as well as hard work, Huawei Electric grew to be the main supplier of a host of power supply products. These were used in China's energy industry, to which Huawei Electric became a primary supplier. In 2000, the company had 1,600 employees and annual sales revenue of RMB 2.65 billion.

In early 2000, anticipating the impending winter, Ren explored the idea of taking this company public under the name of Huawei Electric. Huawei itself held control shares, but more than 60 percent of employees were shareholders. This exceeded the upper limit as mandated by national policy, a rule that said that no more than 200 shareholders could hold shares if the company was to be listed. The plan to list therefore did not go as anticipated. Ren then decided to spin off noncore assets in a way commonly used elsewhere in the world. He reorganized Huawei Electric, divested it, and used the proceeds as a cash buffer to guard against future problems.

The U.S. company Aimosheng is a World 500 firm that has been in existence for more than 100 years. It is the leader in the global core network energy field. Given that Huawei Electric's business results were outstanding in China because of its dealings with China Telecom, China Electric Power, and other high-end power source markets, in discussions with the company, Aimosheng expressed a willingness to purchase Huawei Electric outright for USD 750 million. This astonished senior management inside Huawei, including Ren. Both parties soon came to an agreement. After going through the necessary review and approval process of various Chinese departments at the national level, on October 21, 2001, the two sides formally concluded the transaction. Once the USD 750 million was in Huawei's account, the company deducted the requisite amount in national taxes as well as dividends for employees of Huawei Electric and found it had earned net income of close to USD 500 million.

This was the largest transaction for intellectual property rights in the history of *minying* enterprises in China. Not only did it make Ren Zhengfei recognize the opportunities in what are called "capital operations," but also it quite unexpectedly gave Huawei a new model for fundraising. Starting in early 2003, Huawei spun off various other noncore operations and turned them into either independent or joint-venture companies. When cash flow became a problem, Huawei either sold product lines or transferred rights to the shares in order to directly turn them into cash.

The Digital Telecom product line can serve as an example. H3C was a joint venture set up between Huawei and the American company 3Com in 2003. Huawei maintained 51 percent ownership of shares by putting in pure technology, while 3Com put in USD 160 million and held 49 percent of the shares. In 2006, Huawei ran into cash flow problems owing to its expansion overseas. It sold its 51 percent share in H3C to 3Com, earning a net income of USD 908 million. This was enormously helpful to Huawei's rise in international markets, and it became a classic model for how to raise funds.

In the end, the sale of Huawei Electric not only changed Huawei's intent to list on the market, but also had a much deeper historic significance for the company. Before winter truly gripped Huawei, Ren already had USD 500 million with which to make that nice warm cotton-padded coat. It was impossible to predict how long and how cold winter might be. At least for the next two years of minimal sales, Huawei at least knew it would survive the crisis.

With the comfort of knowing that "there was going to be enough grain to eat and they need not worry," Ren then set out a series of other major initiatives to save the day.

Mending the Fold After the Sheep Have Gone

Ren's strategic decision to put key funding and primary efforts on the then-revolutionary 3G technology was correct, as seen from today's perspective, as were his high-pressure tactics in achieving that end. That is why Huawei submitted more international patents in terms of quantity than anyone on the globe and indeed emerged as a force during the 3G era. At the time, however, nobody could have predicted that the market for Xiaolingtong would become so big and go on for so long. Nobody also could have predicted that the competition would pose such a severe threat to Huawei.

Zhongxing was one example. In the three years from 2001 to 2003, Xiaolingtong sales volume was RMB 2.4 billion, 4 billion, and 8 billion, respectively. Zhongxing had always been behind Huawei, but now in terms of sales alone it began to close in. Moreover, the Xiaolingtong sales volume of a company that was unknown just a few years ago, UTStarcom, now surged from RMB 5 billion in 2001 to RMB 10 billion in 2003. Relying on the high-margin profits of Xiaolingtong, UTStarcom now declared both that it was taking its forces into the fiber-optic network business and that it was beginning to get its feet wet in 3G. All this put unprecedented pressure on Ren.

Huawei had to put an immediate stop to the life cycle of Xiaolingtong by speeding up the process of having the Chinese government issue 3G licenses. Otherwise, once the Xiaolingtong market had become hugely profitable, Zhongxing and UTStarcom would take advantage of their position to extend into 3G. They would further intensify Huawei's crisis by moving into its territory in terms of market position and customer relations. In line with its usual spirit, Huawei decided, therefore, to stick to its high-pressure mantra: "Either do it, or don't do it. If you do it, go all the way. Concentrate talent, material forces, and financial strength to make a breakthrough." As it happened, this enabled Huawei to move into the frontlines of technology after having been a follower. It also prepared the ground for Huawei's cell phones, so that they could later take a prime position in international markets.

In the second half of 2003, Ren spun off the end-user product line from the wireless product division. He then invested RMB 760 million in a new Huawei End-user Company, Ltd. At the same time, Huawei's R&D department swiftly assembled a group of 300 developers experienced in both hardware and software, who began doing "closed-door" R&D in the Hantang Building.

By making this move into the end-user arena, Huawei built on its previous experience and contacts. It moved directly into high-end cell phone core silicon-chip fields through a joint venture it had already set up with Matsushita and NEC. In addition, it cooperated with Japan's Kyocera in Xiaolingtong R&D. By going through OEMs, that is, contracting with others for production, it upgraded the technological content of Xiaolingtong and also greatly shortened the time it took to go to market.

Moreover, to strengthen the look and feel of the product, Huawei selected several Japanese companies that had long been famous for their design of cell phones. He took in the most advanced design concepts in the world at the time for GSM and CDMA cell phones. He quickly developed 12 Xiaolingtong products that covered the high, middle, and low ends of the market for fashion as well as top functionality.

Once fourth-quarter products were put into the market in 2003, they quickly took in 10 percent of the market share for Xiaolingtong, exceeding RMB 200 million in sales volume. This was after just six months of development. Huawei priced its products at RMB 10 lower than the comparable products of UTStarcom and Zhongxing, even though its functionality and style were better. By 2004, Huawei was selling more than 3 million Xiaolingtong phones. Its products had become one of the three legs of the stool for this market together with the products of UTStarcom and Zhongxing.

UTStarcom and Zhongxing had not expected that Huawei would suddenly force its way into their market territory and, indeed, have products with such high functionality-to-price ratio. Their market situation suddenly turned south. Not only did this affect the results of UTStarcom, but also it forced Zhongxing's revenues to start slipping.

Huawei did not actually make a profit out of the Xiaolingtong business. Its pricing policy allowed it to complete its attack on two primary competitors, however, while also giving it time and experience to develop 3G phones. This was critical for the stage to come. After eight years of innovations and constant improvements, Huawei surpassed not only UTStarcom and Zhongxing, but also Nokia, Sony Ericsson, Motorola, and other established end-user brands. It now shared global markets for the end-user cell phone market with two others, Apple and Samsung.

"For those who are ambitious, there will always be opportunities. The lament about being 'too late' is always the excuse of those who are incompetent." This statement by Ren was confirmed by the way in which Huawei turned around the Xiaolingtong situation. The sheep had left the fold, but were retrieved in time.

Different Paths to Survival

In the second half of 2001, as the telecom market moved toward winter, the broadband market was instead welcoming spring. Cisco rapidly took over the enterprise network market in China, based on its powerful strength as the market leader. This was a market worth tens of millions of dollars. In 2000, Cisco's sales volume in China reached USD 500 million, and in 2001, it rose dramatically to USD 1 billion.

With such a market, Huawei would easily survive its own winter. As a result, the company quickly assembled some 100 R&D experts who were soon working night and day on an all-out assault. Within half a year, they had developed a high-end digital telecom product. However, breaking into the enterprise network market and going up against Cisco as an equal first required setting up distribution channels throughout the country that were "seamless," impenetrable. It meant doing this as fast as possible. This was a major challenge for Huawei, which had always used a direct-sales method. Setting up distribution channels was not difficult, but creating a highly loyal team of agents required having powerful brand influence and appeal, but also time.

At this critical juncture, Ren turned this major mission over to Zheng Baoyong. After taking on the responsibility, Zheng Baoyong rapidly formulated a series of completely new policies regarding open recruitment of highly regarded agents. Moreover, he expanded cooperation with several professional agents already working within the country. In February 2002, Baoyong suddenly became ill and had to fly to the United States for treatment. Because of this, Ren now appointed the "top general" Zheng Shusheng to take over the task of opening up the enterprise network market.

After taking on this crisis-laden task, Zheng Shusheng worked hard to develop excellent relations with several hundred professional agent channels. He galloped through the ranks, doing all he could, but Huawei's unfavorable situation made it hard to turn around the situation. One gets to the heart of the matter only when things get really tough—Ren had four standards by which he evaluated whether a manager was worth her or his salt: work ethic, willingness to sacrifice oneself, sense of responsibility, and sense of mission. All four were fully expressed in Zheng Shusheng. After repeated setbacks and multiple mistakes, Huawei finally found its way to a breakthrough point in the enterprise network market by sticking to these standards and sticking to the principle of patience and perseverance.

In March 2003, Huawei and Cisco's old adversary, America's 3Com, set up the company called H3C. The chairman of the board of 3Com served as chairman of this company as well, while Ren served as CEO, and Zheng Shusheng was chief operating officer, now responsible for all operations of H3C. Huawei believed deeply in team cooperation and was familiar with the Chinese situation. 3Com was strong on management and also had 50,000 agents around the world, so it had a wealth of experience in handling distribution channels. The two sides therefore joined their forces as both faced mounting problems in the enterprise network market. They meticulously analyzed Cisco's channels policies, which was an inverted pyramid model. Then H3C adopted a targeted approach of an "upright" pyramid model and made sure that greater profits stayed with the agents.

In addition to this shift, H3C provided agents with all-round support in terms of technology and product promotion. Technological aspects of high-end router products were extremely complex. Because of this, agents often had a hard time communicating effectively with clients. To resolve this, H3C recruited more than 2,000 engineers and distributed them throughout the country to go with agents to explain the technology and to provide solutions that would satisfy customer needs.

To motivate agents, in 2003, H3C promoted several channel policies that were well funded, with names such as the "Sunshine foundation plan" and the "Valiant warrior plan." These provided very effective incentives to agents. By the end of 2003, not only had the company developed a network of more than 10,000 agents of its own, but also it had lured over "gold-class" agents from Cisco, who eagerly joined the ranks of the H3C regiment.

In 2004, H3C realized income of USD 148 million. By this point, Huawei, which had mainly sold via direct sales, now successfully surmounted a host of problems and began selling through distribution channels. Huawei itself earned little income from this. The fact that the effort had succeeded, however, showed that Huawei could indeed find ways to survive. Unlike other companies, it was engaged solely in the telecom industry and had not diversified, but it managed to live on through the bitter cold.

Set Your Heart on Fire

Despite the extraordinary precautions that Ren took to prepare for winter, bitter blasts swept mercilessly across Huawei in the second half of 2002.

After 10 years of superlative growth, the Chinese telecom market saw a precipitous drop in the amount of investment being put into infrastructure. Funding of new assets dropped from an average annual growth rate of 20 percent to 2 percent. Huawei too now began to grow at a glacial pace. Not only did sales revenue show negative growth for the first time since the company was founded, but also the rate at which money turned over in the company, the "circulation rate of money," slowed to a standstill. Huawei was put in a tough cash flow position since it was still investing RMB 5 billion in 3G research even though it was still uncertain to whom the license for this technology within China would go. The company was also continuing to invest in overseas expansion, to the tune of RMB 10 billion. Money was flowing out and not yet flowing in. As a true entrepreneur, however, the more he was in a tight spot, the more Ren emphasized on the bright prospects ahead.

On July 7, 2002, he made an important speech at the third-quarter joint meeting of Huawei's R&D committee and its marketing department. This was entitled, "Understanding the Objective Laws of Piloting the Organization and Making Full Use of the Core Team." In this, he frankly admitted that Huawei had run into unprecedented problems.

However, these were not problems that applied solely to the company but instead involved the wintertime of the entire global IT industry. Not a single operator or telecom equipment maker was lucky enough to escape this, whether inside or outside China. The global nature of this disaster was allowing Huawei to see through to the essence of what lay behind the bursting of the bubble, to adjust its strategies and get up to speed with changes in the global tide.

In this speech, Ren described what he saw as the major trends in the IT industry. After its peak in the year 2000, it was gradually turning into a more traditional business, which meant that battles for existence would now compete more on the basis of quality, service, and costs. In contrast to western companies, Huawei had an absolute advantage in costs. In R&D in particular, its costs were around one-third those of western companies. Not only was it more likely to be able to survive the winter better than western companies on this basis alone, but also it had the advantage over smaller companies because of this. Faced with the current problems, therefore, what Huawei had to do was to reorganize its forces and get ready for future growth.

Ren was highly encouraging. "It is precisely because of our understanding and confidence in our success that we dare to think we can figure things out." His mission now was to train a superlative team that could become even more disciplined by confronting this immediate hardship, so as to be able to deal with what lay ahead. The more his senior managers were roused to do battle, the more they could deal with crises and the greater their confidence would be as they firmly took over markets.

"What do generals do when the field is in disarray and no one can see through the fog of battle? They light beacons by which to lead their troops. As in the Greek myth, they set their very hearts on fire so that their men can see the way."

"The tougher it gets, the more our senior management must light the fire of life itself in the darkness. They must take the initiative, make their troops believe that victory is inevitable, lead their troops toward that victory. All leaders must be like Danco, rouse the spirit of their employees in the depths of darkness, and rouse their fighting spirit and valor. After the battle cry has sounded and the march moves forward, Huawei will be left standing and alive."

After this meeting with senior staff, Ren held a series of meetings with employees to address the problem of insufficient confidence and "fighting spirit." His topics were given such titles as "To meet the challenges, train up your inner *qigong*, and get ready to meet the springtime that must follow." He told everyone that when the market was easy and obliging, it was simple for everyone. When the market turned tough, that was when the heroes among people stood out and showed their true colors. He sincerely hoped that employees would "fan up" the inspiration of the old days, stiffen their spines, and get through these next few years of hard work.

"When is morale the best? If morale is only good when the market is red hot, but bad when things get tough, who will we have to turn hard times into future growth?"

"The tougher the times, the more you can forge ambition in the hearts of your people, train a moral and conscientious approach in them, raise their skills and abilities. Indeed, this is the most important time for us in building up our troops. The best time for developing a positive corporate culture and a team spirit is *not* when everything is going smoothly. Instead, it is when the company is facing setbacks. As the ancients said, people show their true colors when faced with adversity. This follows the same principle."

In motivating his team and encouraging a "fighting spirit," Ren gave examples of the most famous battles in China as well as around the world, in both modern and ancient times. At decisive moments in such campaigns, the formation of troops, the "organization," did not unravel. This was the basis for ultimate victory. Similarly, in telecom, since every competitor in the world was facing the same situation, Huawei had to maintain its discipline and not become chaotic. If it was able to do this, it would achieve tremendous success in the next two to three years in international markets. What's more, hope was already within sight.

Finally, Ren would always present the main reason he had brought the entire group of employees together in this kind of meeting. "Everyone is asking, 'should we aim for the World 500?' What I say is that we should put an end to this very idea, from the top of the company down to the bottom, give it up. We should never say that 'it will not be one, but two, three generations before we get into the World 500.' And when

I say 'generations,' I'm not saying one or two generations of leadership, but instead one or two 'rebirths' of the company, from being down to pulling ourselves back up again, and again, and again. Instead, our definition must be that 'we are going to survive.' If we define it that way, we will indeed survive."

Leading the Way in Setting an Example,
Being in the Same Boat in Stormy Weather

Such courage and determination in the face of a hopeless situation profoundly affected Huawei people. Ren's honesty and determination were infectious. After the meeting, some key leaders in the finance department led the way in proposing to the human resources department that salaries be reduced. After the New Year's vacation in early 2003, the first day employees came back to work, Ren, Sun Yafang, and senior management at the CEO level voluntarily lowered their own salaries by 10 percent. In what was called the "454 document," they applied to the human resources department for lower pay. After this, cadres at the "ke" or division level and above responded in similar fashion, expressing their willingness to help the company get across a tough patch by taking lower pay. They would all be in the same boat going through the storm.

"This reminded us all of the behavior of people during the Asian financial crisis in places like Japan and South Korea, where people voluntarily brought out their own belongings to contribute to the country in order to help it get by, their jewelry, their cash, anything they had." Remembering those times, one of Huawei's managers actually became quite emotional.

Since managers had led the way in setting an example, employees now took action too. Department by department, they declared their willingness to fight on behalf of the company. They instituted austerity measures to carry the company through. Departments fought to be first in declaring this intent in the *Huawei People* newspaper, the *Better Management* journal, and on the company's internal website.

As an example of the new frugality, expenses were cut back. Huawei had spent lavishly on corporate meetings when things had been going well the previous year. Each year, there were countless meetings of various sizes, for training sessions, exams, corporate gatherings, and so on, given the thousands of employees in the company. These had always been held in three-star hotels or above. Reimbursement for travel costs, meals, meeting costs, and so on had been extremely high. Now such meetings were held in a virtual format as much as possible and via the internal network. This saved time and also a tremendous amount of money. People now printed on both sides of the paper, used less water, and turned out lights to cut costs.

Beyond this, however, out of respect for Ren and out of a real love for Huawei, the great majority of employees showed the kind of loyalty that is hard to come by. Specifically, they turned down offers from other companies. For example, the head of sales in a particular department got calls from headhunters nearly every day in the first half of 2003. They enticed him with offers that were several times his salary at Huawei. Competitors even came to him with major financial inducements to give up certain critical orders. He refused these in no uncertain terms.

"European and American companies have been firing large numbers of employees. In contrast, we have kept our company whole. Other than our loyalty, what more can we contribute to Huawei?" Such statements seemed humble and simple, but they truly did express the feelings of the majority of Huawei employees at the time.

When a corporation is at a critical stage, either its keepers can provide employees with a sense of hope, or they can allow them to think there is no hope. When the global telecom situation was in fact facing its worst situation ever, many companies fired personnel as a way to survive. Ren did not fire people, but what is more, he found ways to allow employees to see a glimmer of hope. They could see spring ahead in the depths of winter. It was this entrepreneurial acumen and the quality of the man that enabled Huawei to emerge from its dire straits.

Ren wrote movingly of this time as he recalled it in an article called "The Spring Thaw Flows Toward the East." "We propped up the company by deciding collectively to lower salaries. We made up for the mistakes of our youth in a kind of selfless approach to work. We put cotton padding in our winter coats by opening up overseas territories, sending people off away from families and friends to get it done. From top to bottom, the company worked with one mind and one intent, we 'tasted the gall in order to stay wide awake,' which enabled us to reach where we are today."

The Movement to Change Share Rights

In the midst of hardship, Huawei was able to get employees to stay loyal and fight with unbending spirit. As things continued to worsen, however, and winter became colder, this approach might not always work, so other measures were also necessary. In order to reduce operating risk as much as possible, Huawei now began to set up institutionalized safeguards. Ren led the way in instituting a new shareholding system that would define the company's ongoing ability to exist.

The shareholding system for employees had begun in 1992. At that time, Huawei was in a tight situation since all money was being put into research and development. The system Ren then adopted played a decisive role in motivating employees and attracting new talent. He called it "bearing the burden of corporate risk together, and sharing in corporate success together." This then became a kind of "filling station" that could be resorted to if needed to spur high-speed growth. However, as the company hired new people at a fast pace, there began to be a considerable difference between the numbers of shares held by older and newer employees.

In order to resolve a problem that was the legacy of corporate history, on September 15, 2003, Huawei announced a management buyout plan to all employees. This would include a total of 1 billion shares. Understanding that newer employees would not have that much money to spend on shares, Huawei decided that they would have to pay in only 15 percent of the cost. Huawei would itself borrow the rest from a bank, in the name of each individual employee. Moreover, shares purchased in this way could not be transferred, sold, or used as collateral in any way for a period of three years. There was therefore a "lockup period" of three years on this allocation of shares.

After this management buyout plan was announced, most new employees were enthusiastic and eager to participate. They quickly signed share-purchase agreements. After the National Day celebrations in 2003, the finance department began undertaking the relevant procedures for loans from banks.

This initiative effectively reduced the disparity in shareholdings between older and newer employees. It provided an enormous stimulus to new employees and got them to maintain enthusiasm and a sense of responsibility about their work. Meanwhile, the three-year lockup period served as an institutionalized check on employee departures and it stabilized "the psychology of the troops." It tied the fate of employees more closely to the fate of the company. It served as a critical "weight" in balancing things out within the company and allowed Huawei to come through the crisis safely.

Counterattack When up Against the Wall

Between 2001 and 2003, as Huawei went through the coldest part of winter, many employees did indeed leave the company and take corporate secrets, codes, and technology with them. Once these flowed outside the confines of the company, all sorts of low-cost products began to appear in the market since the products had involved zero investment in R&D. This served as a potentially mortal attack on Huawei.

Huawei took every possible countermeasure to prevent the stealing of technical secrets and classified information. For example, it installed surveillance equipment in the offices of every R&D center. It established explicit rules in R&D departments that locked up all USB ports and hard drives in order to keep people from divulging information. It strictly forbade any access to the Internet. If there was a special need to connect via Internet with customers, this had to go through a stringent permitting process and then monitoring by the Information Security Department.

Despite these strict practices and all the security systems, the theft of technical secrets continued to be rampant.

On November 21, 2002, beside West Lake in Hangzhou, in the middle of the night... two employees of UTStarcom were arrested on criminal charges by the Guimusi police authorities with the coordinated assistance of the Hangzhou police. The names of these two were Wang Zhijun and Liu Ning. Another UTStarcom employee named Qin Xuejun was also arrested after getting off an airplane at the Xiaoshan airport. After this, as per proper judicial procedures, the Guimusi police handed the case over to Guangdong for interrogation and investigation. The news of this incident quickly surfaced.

The three people—Wang Zhijun, Liu Ning, and Qin Xuejun—had all entered Huawei after graduating with master's degrees in engineering. They were involved in developing optical transmission technology at the company. On August 9, 2001, the three individually carried out resignation procedures. On November 7, 2001, they established a company in Shanghai called the Hu Ke Company.[1] In October 2002, the UTStarcom company purchased the Hu Ke Company for RMB 2 million in cash and another USD 15 million in options.

Not long afterward, an engineer in customer service at Huawei was working with the telecom bureau in Guimusi when he discovered that the optical transmission equipment that bureau was using was completely compatible with Huawei's equipment. Moreover, the two systems could connect seamlessly with each other. Clearly, somebody had stolen this core technology that belonged to Huawei. After Huawei reported the case to the Guimusi police, the Guimusi authorities reported it on up to the Public Security Bureau, since the case was so unusual. It involved not only a listed company in the United States, namely, UTStarcom, but also Shanghai Bell, the joint venture of France's Alcatel in Shanghai. The Public Security Bureau then became responsible for coordinating efforts to investigate the case among the Hangzhou, Shenzhen, and Guimusi police departments.

As everyone knows, optical fiber transmission technology is one of Huawei's core technologies. In 2002, its global revenues from this technology ranked fourth in the world, but from 2008 up to now, it has ranked first in the world. Over years, Huawei has invested many billions of RMB into the technology, and over 1,000 R&D people have amassed a tremendous volume of research results. If Huawei did not press criminal charges to put a stop to this kind of problem, using civil procedures alone would very likely have dragged on in a kind of legal marathon. Meanwhile the

---

[1] *Hu* is another name for Shanghai, and *ke* refers to technology, so the Hu Ke Company is the Shanghai Technology Company.

technology for optical transmission would more than likely flow out into the public realm. The main concern within the company therefore became to protect the results of its own research. Huawei was already up against the wall, and in fact it could go under at any time. It had no alternative but to counterattack and adopt criminal proceedings. However, what also emerged from this whole case was a more broadly based and quite serious social issue. How was China to use legal processes to protect Chinese intellectual property from being infringed upon?

On December 6, 2004, the critical evidence in the case was handed over to state-level "authoritative institutions" for evaluation. After this, the District Court of Shenzhen's South Mountain[2] district made the following determination: the three defendants had violated an agreement they signed with Huawei to protect corporate secrets. This made Huawei incur massive economic losses, and it therefore constituted the crime of infringing upon corporate secrets. Wang Zhijun and Liu Ning were sentenced to three years of imprisonment,[3] while Qin Xuejun was sentenced to two years' imprisonment. Assets in the Hu Ke Company account that had been frozen were ordered to be paid out to Huawei as partial compensation.

"Future Battles Among Companies Will Be Battles over Patents"

The fanfare over this case died down, but the battles themselves were far from over. Ren now began an incessant publicity campaign, partly to educate employees within the company but also to stir up public awareness as well as government awareness about the need to protect intellectual property and to set up relevant institutions to do this.

On April 11, 2005, Ren was inspecting the Huawei R&D center at Bangalore, in India, when he told senior management there, "One of the reasons India has become a major player in the global software industry is that its government has been extremely forceful in protecting intellectual property. It has adopted stiff measures. The monthly income of Indian

---

[2]Or Nanshan.

[3]The punishment is literally "being put in prison for a specified period of time as opposed to unspecified, i.e., life imprisonment."

people is around USD 50. If it discovers pirated software, such as pirated CDs, however, the government puts on a minimum fine of USD 30,000. The minute it discovers large-scale infringement, whether by a company or an individual, the result is an utter calamity for either the corporation or the individual."

"In China, where piracy is rampant and violation of intellectual property rights is pervasive, we will pay a tragic price for it in the future if we do not quickly turn ourselves into a 'trustworthy society.' If you can invent technology but cannot protect it, you will fall behind and you will be beaten up. This is simply one of the laws of history."

On April 28, 2005, Ren drove home this point in a report delivered at a meeting that he called "Huawei's Core Values." "How is economic globalization defined in its most essential terms? In the past, it was warfare—war is what made economies interact on a global basis. In the 1970s and 1980s, it became industrial manufacturing. In our current age, future competition for markets will be battles fought among companies, which means battles for patents. Any country without its own core patents will never become a strong industrial nation."

In the end, Ren's efforts were widely acknowledged as the Chinese government came out with a succession of policies, laws, and regulations. From a legal standpoint, these strengthened and improved upon judicial protection of intellectual property rights. As a result, they enormously boosted Huawei's own drive to be innovative. In 2006, Huawei applied for 575 patents from the International Patent Office, making it thirteenth in international applications. In 2007, this figure rose to 1,365, and the company jumped to fourth place. In 2008, it applied for 1,737, putting it ahead of Matsushita (with 1,729) and Royal Phillips in Holland (with 1,551) to be the global champion. At the same time, the cumulative number of patents that Huawei had applied for now reached 35,773. This was more than 50 percent of all international patents applied for by Chinese companies. Huawei was leading the way in emerging as a global player on the patent stage.

In the midst of winter, Huawei again came through a difficult patch in its history, but it still had to deal with other issues as it waited for spring.

CHAPTER TEN

PRACTICE INNER *QIGONG* WHILE WAITING FOR SPRING

Winter is not something to be afraid of, or to avoid. Without it, we stay fat and overconfident. Huawei absolutely must not allow itself to be arrogant. For that reason alone, wintertime has its merits.

—REN ZHENGFEI

By preparing a warm cotton-padded coat, mending the sheepfold, and collectively lowering salaries, Ren Zhengfei prepared to get through winter. By undertaking the changes in share ownership and "counterattacking when the back was against the wall," he turned Huawei from unprecedented turbulence in the direction of relative stability. Although Huawei paid a price for this, it did not get mired in debt and swept away, as did such companies as Nortel and Marconi. As the ramifications of the burst IT bubble worsened, that is, as the cold intensified, and the telecom industry around the globe was engulfed in it to one degree or another, Huawei also faced an incomparably good opportunity to change the global pattern of competition.

"How do we obliterate the effects of the bubble? We improve our per capita results."

On July 7, 2002, Ren made a speech at the third-quarter joint meeting of Huawei's R&D committee and its market department entitled, "Understanding the Objective Laws of Mastering Control, and Making Full Use of the Core Team by Constantly Raising per Capita Results." In this, one can hear the overture of his plan to take Huawei out into springtime.

Guiding Change into Deeper Waters

Low per capita results had long been a bottleneck preventing China's enterprises from moving forward. Given his frequent interaction with consultants from IBM and Hay, and his close observation of world-class enterprises, Ren summed up the core lessons in two different respects. First, different parts of the company, different organizations within it, were developing in an unbalanced way. Second, Huawei had long practiced a system that made people responsible to other people, instead of having them be responsible for actual results.

This situation had come about naturally, for historic reasons. During the start-up period in the company, as in other *minying* enterprises in China, Huawei was pressed to survive. It was "waiting for its food every moment." The two pillars supporting its survival were R&D and marketing, which is why the majority of senior managers were chosen from these two departments. When corporate decisions were being made and promotions were being granted, out of habit these managers would lean in the direction of their former departments. They would also neglect such areas as finance, production, planning, and quality control when it came to allocating resources. As a result, the strong departments got stronger and the weak ones became weaker.

As Huawei grew quickly and its operating mechanisms improved, this chronic problem was never dealt with in a rational and scientific way. Instead, it became more serious. For example, when the "dragon's head," the R&D department, constantly made changes in the design of products, the dragon's body was forced to change as well. Everything from production, procurement, and materials, to dispatching of products, was in disarray on down the line. One could also use the analogy of a bucket with different lengths of staves. The amount of water the bucket could hold was determined by the shortest vertical stave that made up its sides. Even though Huawei had cutting-edge technology that was on a par with international standards, its "shortest stave" determined its overall results.

Meanwhile, a great deal of redundant labor led to low per capita results, so that what should have been the advantage of low-cost labor in China actually was not. Huawei's overall competitiveness was still fairly low.

Ren pointed this out quite sharply in internal meetings. "If our system of performance evaluations does not bring departments into greater balance, some departments will continue to lack outstanding cadres. They won't be able to keep up at the same pace. And if they don't, it is patently obvious that overall results won't get better. This kind of lopsidedness means that it will be futile to talk about making progress."

Balancing Forces

In July 2002, under the guidance of IBM's consultants, the supply-chain systems of departments within Huawei were revised according to world-class enterprise models. These departments included production, quality control, procurement, warehousing, and logistics.

In the production and quality control departments, Huawei hired qualified authorities who had retired from the German National Applied R&D Institute. They were paid high salaries and asked to come reorganize the company's entire production system. One quite obvious change that they made was that all product cycles and product quality standards were now judged according to international standards. Once a product entered the production line, it was not allowed to be changed in any sudden way. This gradually standardized the rate at which products, quality control, and orders were consistent.

In the logistics department, since suppliers of materials and equipment suppliers for Huawei came from all over the globe, the company dealt with tens of thousands of different raw materials, semifinished products, and categories of products. Huawei's logistical department therefore had more than 2,000 people at headquarters alone, but these people were not effectively managed. Materials were not supplied on time for production, and materials were sent to the wrong place or the wrong materials were supplied. With the help of the German experts, Huawei's production base at Bantian now set up a fully automated first-class warehouse. It may have seemed as intricate as a spider's web, but in fact it was highly organized. The previous 2,000 people were now reduced to 35. This not only lowered wage costs and warehousing costs, but also ensured that goods were supplied on time. It proved enormously helpful in safeguarding production schedules.

Procurement was another major problem in Huawei. One vice president involved in this side of things had the following to say during a procurement meeting: "Back when Huawei was quite profitable, it paid little attention to cost. It would often choose the most expensive when it bought full sets of equipment or instrumentation. It would then brag that it had the most advanced in all of Asia or even in the world. Not only was this enormously wasteful but the equipment was hardly used."

In 2002, Huawei hired a retired vice president from IBM to serve as head of its procurement department. He was paid an annual salary of USD 600,000. His job was to turn Huawei from a small workshop into a company with a world-class procurement system. Any purchase of any kind had to be done after a publicly announced tender, and it had to be sent to global suppliers according to established international practices. Then, after a "price relative to performance" evaluation, the procurement department had to undertake a detailed financial and functional analysis before coming to a comprehensive determination. At the lowest possible amount of financial outlay, it would purchase what was most needed and could most be used. In 2002, Huawei was able to save more than RMB 2 billion on one single purchase alone.

At a meeting regarding this subject, Ren declared, "Competition among enterprises these days is not simply a matter of one enterprise going up against another. Instead, it involves one supply chain going up against another. A company's supply chain is like the entire chain of an ecosystem. It has all of the lives onboard the ship, including customers, partners, suppliers, and manufacturers. A company can only expect to survive over time if it also looks after the interests of its customers and its partners, and if it pursues win-win solutions for all."

Facts were to prove that Huawei's 27 years of growth, its long-term accumulation of competitive advantage, was not so much a matter of core technology as it was a matter of low-cost advantage. Lower costs are what gave Huawei the ability to compete on an all-around basis with multinationals. Reform of the company's supply chain, by addressing the "short staves in the bucket," was critical at this point in the company's history. It enabled Huawei to move forward in a more balanced way.

Using IT to Promote Management Advances

Ren believed that another reason for the low per capita results at Huawei was that it paid fairly high wages and benefits. This made many cadres "play it safe." Afraid of losing their own status, they constantly asked for instructions from above and then implemented management's advice in a dogmatic way. Over a long period, this then led to a system of "being responsible to people" instead of "being responsible for affairs."

"Huawei still operates in a small workshop manner. It is not yet accustomed to professional, standardized management that operates by certain models and metrics. We still have redundant layers of management and unnecessary work being done, which is the root source of Huawei's inefficiency."

"The things keeping us from pushing forward IT are mainly coming from inside. Senior and middle levels of management are afraid that they will lose authority once we implement IT systems in management. But I ask you, don't we all know that the company's very existence depends on improving our management systems? Such improvement means getting rid of many intermediate links, speeding up end-to-end and point-to-point processes, and making them more accurate."

This was Ren's conclusion in April 2001, after the investigation trip to Japan. He analyzed Huawei's own management deficiencies in the report referred to above called "Springtime in the North." In a previous article, published in the 65th edition of *Better Management*, he also elevated the issue of IT building to the status of "an issue that will decide Huawei's survival." In this, he wrote about "using IT to propel advances in management." Shortly after that "wintertime" arrived, however, and he had no choice but to slow down this initiative. Now, as the crisis seemed to be abating, it again became a major focus for the company after 2003. The issue became how to use information technologies to propel advances in management, raise per capita results, and make people responsible for results as opposed to being responsible just to their superiors.

Back in 1998, Huawei had spent a large sum of money to adopt a U.S. online office system called "Notes." Since management reform had not yet been launched at that time, however, this system was used to minimal effect. It was used to issue notices, download office documents, and so on, as a kind of paperless office instead of using faxes and the mail. As the "eight large management reform projects" were carried forward, however, including IPD (integrated product development), ISC (integrated supply chain), "unifying four financial aspects," and so on, basically all operations of the company were now incorporated into this system. The project management department gradually set up a functioning IT management platform and enabled Huawei to undertake a fundamental transformation in its online office procedures.

As one example, Huawei was now able to issue new versions of product technology at a dramatically increased pace. Before this time, the company had used printed materials and face-to-face methods of promoting new products to customers. Now customers and partners were authorized to enter into Huawei's IT management platform to understand and learn about new Huawei products themselves. They were given access to training on new technologies. Not only did this save time, but also it greatly improved work efficiency.

As another example, the problem of reimbursing for expenses had always been the cause of inaccuracies in estimating Huawei's costs. Since Huawei's branch structure covered the entire globe and had a tremendous number of employees, the frequency, extent, and timing of expenses were all different. Some employees would submit expenses only once every six months. Some expenses would come in only once every two years—those related to certain provinces within China, such as Xinjiang and Qinghai, as well as in some international offices. Apparent costs and hidden costs were therefore highly uncertain, which led directly to overestimating as well as underestimating actual expenditures. This made it impossible for real profitability to be reflected in current figures. It also made it hard to evaluate the contribution that each department had made to results with any degree of accuracy.

Now, a system called "electronic reimbursement reports" was instituted as a part of the IT management platform. This changed the situation considerably. Under strict regulations issued by the finance department, several tens of thousands of employees now were required to send in reimbursement reports monthly, with costs divided into specific categories. This made the true nature of costs and profitability float to the surface. It enabled management to make appropriate policy decisions with respect to performance, taxes, investment, and cost controls. Moreover, auditing departments could now access the IT management platform and conduct oversight on the process of investing in R&D projects. They could conduct audits of costs in various departments, including production, procurement, warehousing, and supply-chain systems. Huawei's previously chaotic financial management system became far more transparent and orderly.

By the end of 2003, through constant reorganizing and improving, Huawei had set up a fully equipped modern IT management platform. This lowered management costs dramatically and cut down on superfluous costs, and it notably improved per capita results and per capita efficiency.

In 2000, Nokia's per capita results came to USD 440,000, and Nortel's came to USD 320,000. Huawei's, in the same period, came to USD 130,000. That is, it was 2.9 times worse than the average of these two companies. By the end of 2003, Huawei's average was 2.2 times worse—that is, despite three years of effort, the disparity had not been reduced by much. However, to Huawei this represented an enormous advance. Due to Huawei's innate advantage in low-cost labor, and the fact that its "working time" was around twice that of these two companies, in terms of per capita results, Huawei did in fact begin to approach the baseline levels of multinationals.

Collective Decision Making and Separation of Powers

With the full-scale implementation of core management processes that centered on IT systems, including such things as IPD, ISC, and the "unification of four financial systems," Huawei began to have the basic "hardware" to become a multinational company. However, an important consideration remained the establishment of sound internal policy-making mechanisms and constraint mechanisms, if the company intended to become a lion as opposed to a mere hyena.

From 1988 to 2000, Huawei went from being a small operation with only six people to a large-scale "group" enterprise with more than 20,000 employees. During this time, Ren used his personal charisma and strategic vision to discover opportunities, seize them, and command his "three armies" to advance, thereby leading to the swift growth of the company.

After this period of high-speed growth, the structure of market competition and the environment for ongoing growth in the industry became increasingly uncertain, even as Huawei continued to push out new products and enter into new business areas, including its international campaigns. As the reform of Huawei's management began to show promise in 2003, therefore, Ren now explored changes in the company's organizational structure in a more concerted way. He did this in partnership with IBM consultants. The aim was to lower operating risk to the greatest extent possible and to create solid ground for Huawei as a multinational in the true sense. The idea was to follow rational procedures as the company evolved.

After five years of close cooperation, IBM consultants were by now extremely familiar with Huawei. They pointed out that a multinational should possess a sound "central axis" organization. They felt that Huawei's senior management positions were instead "hollow." Titles were there, but the people did not in fact participate in major strategic decisions. Huawei relied excessively on an extremely small number of "heroes," and these heroes could not be replicated. The consultants therefore recommended that Huawei adopt IBM's decision-making model by implementing an EMT (Executive Management Team) organizational change.

After considerable reflection and weighing of options, under the guidance of the IBM consultants, Huawei set up seven major departments: "market and services," "strategy and sales," "products and solutions," "operations and payment," "finance," "tactics and cooperation," and "human resources." At the same time, it eliminated the positions of "standing vice presidents" that Huawei had used for many years. It turned these into positions of senior decision-making authority in the company.

Ren took the lead in agreeing that the company needed to avoid the risk that was involved in having one person make all the decisions.

After more than half a year of operating this new system, Ren discovered that it was in fact very difficult for people to cast off the massive influence of a system to which they had grown accustomed. What's more, they had no experience in running a large corporation. As a result, when it came to making decisions and discussing issues, what appeared to be collective decision making was in fact merely a reflection of each person's own ambition.

To push EMT teams away from reliance on him and to realize greater team "value requirements" and ambitions, to make them have a more global perspective and international thinking, in 2004, Ren decided to implement a rotating chairman system within the EMT team. Each of seven members of the EMT team would become chairman of the company as a whole on a rotating basis, with a term of six months.

Implementing this rotating chairman system not only strengthened internal oversight and enhanced a climate of competition, but also balanced out conflicts that had existed among the company's various departments. What's more, during the time that each chairman was in office, he had to tend to massive amounts of daily affairs but he also had to prepare and draft documents for senior management meetings, to formulate growth plans for the company. This was extremely helpful in building up management capacity and decision-making ability in the company. At the same time, to gain support for any action, each chairman had to consider the interests of others. The rotating system unavoidably whittled away each new chairman's own "rear end," his individual support system. That brought his own department into balance with the overall interests of the company. All departments therefore began to coordinate their efforts with one another, to cooperate by uniting efforts. High walls that had been erected over a period of years among departments, and the forest of so-called "peaks," gradually began to erode as the rotating chairman system took hold.

To ensure that the "central spine of Huawei's nervous system" did in fact move in standardized ways, Ren instituted another major change in organizational structure. Any department that was grade 3 or above had to institute three different systems: office committees, administrative management teams, and "commissions." Each of these three forms of organization had its own authority, "the right to propose," "the right to grant approval," and "the right to reject." The core idea of this was that, during any cadre's term in office, the head of the department had the right to suggest who would be promoted but not the authority to grant approval for that promotion. The administrative management team had the authority to grant permission but not to suggest names. The "commission" mainly was responsible for testing a given cadre's qualifications and capability. It had the authority to reject a proposal. It had the right of refusal.

In an internal meeting, Ren provided an explanation for this system. "We have set up a system that splits the authority to make promotions into three separate parts. The business departments can propose names for promotion, the human resources system can evaluate, while the commission system has the right to reject. With respect to how we evaluate cadres' performance, we also have three different requirements. The first is that responsibilities must be results-oriented. This requires performance evaluation mechanisms that judge a person according to key job-related matters. The second is based on the job descriptions in the company's strategic levels and grades. This involves the company's system of 'washing out the last in line,' the one who performs least well, as well as the company's promise to each individual in terms of performance-based compensation. The third is based on the jobs at each level which are evaluated according to certification procedures and mechanisms that promote professional skills. In this process of evaluating cadre performance, we do not focus solely on results, because those may show just that you are trying to avoid being washed out, while they may not show that you are qualified for being promoted. The key thing in this system is to look at a person's overall behavior and judge on the basis of comprehensive factors."

This system had the effect of setting up "clean government" among cadre teams, since it divided authority into three components and set up internal constraints. It also was critical in preventing the further spread of such things as "creating factions by pulling in your own friends" and "only appointing or promoting your own relatives." The aim was to safeguard the practice of allowing outstanding people to rise to the top.

At the same time, however, more intense competition for positions within the company furthered a vivid awareness that "one can go up but one can also go down."

Promoting Professionalism

After the core institutions were set up in the company, Ren's next major task became to figure out how to make more than 30,000 employees all row in rhythm. This meant promoting professionalism in the organization.

Huawei was born into the midst of international competition that existed at its own doorstep. It had to deal with giants in the global telecom community. The company's "culture, spirit, and beliefs" were major pillars that supported its constant growth as a "people-operated," or *minying,* company. "In victory, we raise a toast to one another, in defeat we fight to the death to save one another." "The brave one wins the fight on a narrow road." "We fight for the rebirth of the Chinese nation and for the well-being of our families as well as ourselves." Such tenets could whip up an absolute fever in employees. Such articles of faith seeped into the very bloodstream of Huawei people, especially after the company achieved major success. They became "totems" of the Huawei spirit. In trying to bring management in line with international practices, therefore, Ren creatively brought forth three rather famous "conclusions" with respect to making Huawei more international, more professional, and more mature.

"We will become international only if we break through our narrow sense of self-regard as a nation. We will only become professional if we break out of our sense of arrogance as a company. And we will only become mature if we break out of the narrow way in which we conceive a 'brand.'"

He set forth these conclusions in an internal talk that was given the title, "The Responsibility and the Mission of a Professional Manager." He talked about Huawei's historic mission to become professional, and he defined this and positioned it quite explicitly. "Huawei was a small company created by so-called heroes that is in the process of evolving into a company of a certain size with professional management. It is now necessary to mute the colors of the heroes, particularly senior management and the founders, if we are to realize professional management. We can improve operating efficiencies and lower internal management costs only if we institute regular procedures. One major attribute of the 'second founding' of a company is that heroes now find it hard to rise to senior management levels, and even to survive."

He continued, "Each professional manager operates within a standard portion of overall procedures. I liken it to a train that is going from Guangzhou to Beijing—hundreds of people help in switching tracks along the way, and dozens of drivers take over from one another. You can't say that the driver who finally pulls the train into Beijing did all the work. Maybe one person has to go get the roses and the glory, but he represents the whole group. He is not truly a hero."

In addition to guiding employees in this kind of correct thinking and methodology, Ren undertook a major move to make sure that a professional approach really did take root in Huawei. He instituted what was called the "Five-grade double-track system." He did this with the assistance of the Hay Management Consulting Company, which integrated England's job qualifications system with Huawei's unique characteristics. The system provided Huawei employees with two possible development paths in their professional lives, one on the technology side and one in management. On the technical path, five different grades were defined: assistant engineer, engineer, high-level engineer, technical expert, and highly qualified technical expert. The management path was divided into three levels of supervision, four levels of management, and five levels of leadership. Each employee could design his own career path depending on his own personal interests, qualifications, and expertise.

For example, if a person in research and development set his sights on accomplishing things in the technical sphere, depending on his job performance as a technical expert, he could earn the same compensation as a person in management who was at the same level. If an employee felt that his personal capacities were very strong, he could aim to further himself in both management and technology at the same pace and then later decide what he ultimately wanted his job position to be, depending on the results of his performance evaluation.

This way of doing things helped outstanding people come to the fore. At the same time, it created an environment of equal competition in which

people could do what they felt best at and really wanted to do. It enabled
the company overall to move toward professionalism at a faster pace.

Living in Peaceful and Rational Times

Collective decision making, the tripartite division of authority, and the
"five ranks with concurrent movement along two paths" provided a stage
on which Huawei people could develop their own unique strengths and
capabilities. The processes became institutionalized, in theory. To make
these institutional innovations a reality, to put Huawei on solid ground
and have it evolve in the direction of being a multinational, Ren used his
personal charisma and ability to set an example. This yet again played a
decisive role in training people to try to come up to his standards as he
served as the rod against which others were measured.

Learning English was one example. In the course of the opening up
of international markets, people's level of English deteriorated because of
the crush of daily work. In contrast, the sound of English coaching issued
daily from Ren's office, and his own level of English quickly improved. He
would even try to put some western-style humor into conversations as he
used the language, though not in work-related situations. The practice at
higher levels quickly gets copied at lower levels. A "study English craze"
soon swept through the company.

Another example involved speaking one's mind. Prior to this time,
Ren's own acumen would make cadres think twice before saying anything in meetings or discussions, or when making decisions. Constant
reinforcement turned this into a kind of reliance on him. Now he did
everything he could to encourage them to be courageous and speak out.
He also began soliciting their views on things and their proposals. As the
rotating chairmanship and EMT policies were launched in particular, one
would often find him saying, "Our decision-making system has changed.
We no longer have an environment in which one person is an autocrat.
Whether you want to or not, you have to hold meetings and talk things
through—just doing that will have an influence. Only if we strengthen
the authority of others to say no will we be able to institute a sound
corporate system that does not rely on any single individual."

In this regard, he emphasized, "In Huawei, I should be the one who least attends to things, who is not concerned with specific matters, and who doesn't try to seize power. My job is to create an environment for others to grow the company. Each leader must learn how to be a leader, create a context in which everyone can fight effectively. Each has to recognize that the success of his own troops is what defines his greatest success."

On May 25, 2003, Ren accepted an invitation to speak at a cadre management training class, and he delivered a speech that he called "Living in Peaceful and Rational Times."

PID, ISC, CMM, job qualifications, performance evaluation systems ... all of these are basically just methodologies. It is only when they seem not to be there at all do you know that they have served their purpose and been alive. Their unobtrusive quality is best expressed when management comes or goes, and yet the management system carries on generation after generation. Their viability is expressed as each generation of "fighters" comes to the end of its tenure and yet generation after generation of managers keep getting more mature.

The reason I push so hard for IPD, and ISC, is that we can get what we want done through fewer layers of administration. We have lower costs and higher efficiency, and we get away from reliance on one individual in the enterprise. We get things done by tying together simpler procedures with greater control, by linking input and output more directly, and by tightening up the whole process from end to end.

The enterprise has reached a much more viable position once an entrepreneur in such an organization no longer has much to do. When an entrepreneur, say, the founder, carries too much prestige, when everyone still worships him, that means the enterprise does not have much hope. It is vulnerable. I believe that the "macro-business model" for Huawei should therefore be that management is oriented toward procedures, that we are a procedurally oriented organization and that we produce products based on customer need. We must keep firmly in mind that the customer is, eternally, the soul of the enterprise.

The speech called "Living in Peaceful and Rational Times" summed up Ren's management reform in Huawei. In it, he set forth the credo by which Huawei would be able to sustain itself over the long haul. With respect to his own role, it "muted the colors" of the entrepreneur and strengthened a more professional management. It sketched out ways for Huawei to be institutionalized. At the same time, it showed how a people-operated, or *minying*, company in China could transition from "rule by people to a rule of law." Because of this, every manager in Huawei also began to shift in his thinking from quantitative to qualitative metrics, in terms of their approach to business. The heads of departments began to put greater effort into ensuring that their departments were more coordinated with other departments, and that the people under their jurisdiction contributed to this shift. Individual employees became much more focused on their own immediate work, and they delivered better results. They all tried to speed up their own promotions by the double-channel approach to the five rankings in the company.

In the course of the IT winter, Huawei employees went through the painful periods of slow and negative growth. They personally witnessed the trajectory of two other IT companies that went from glory days to collapse. They had a clear awareness of the huge risks in the IT industry and therefore had a visceral understanding of Ren's constant admonition to have a "sense of crisis." They handled things in a more low-key and prudent way. As a result, Huawei cadres became much less "wolf-like," as the multinational genes of the company also became more and more pronounced.

Complete Reincarnation: Shucking off the Flesh and Remaking the Bones

Ren was steadfastly determined to cast off "Chinese characteristics" and "Huawei characteristics" as he reformed management in a way that cast off the flesh and remodeled the very bones of the company. After five years of sustained effort, the company was indeed "reborn," whether that was defined in terms of its body, its spirit, or its bones.

The vice president of Huawei's R&D department, He Tingbo, remembers when the IBM consultants delivered one last class to the R&D department, prior to pulling out of Huawei after five years of management reform. They revisited the 10 "diagnoses" made at the start of the project five years earlier. In September 1998, at the first class, IBM's consultants had set forth 10 management problems that Huawei needed to address. Now, those listening to the final class were astonished to find that 9 of the 10 had been resolved to the point that they were regarded as plain common sense. IPD had finally been absorbed into the bloodstream of Huawei people. It had completely changed the way Huawei operated.

The company now operated with only one goal, which was to satisfy customer demand and, in so doing, earn fast profits. From the first day of developing a product, to marketing, to finance, to R&D, to service support, all responsibilities and roles were now incorporated in an overall process, and each had its corresponding accountability when it came to finance.

Another large reform, called integrated supply chain (ISC), was even more apparent within Huawei and contributed to its core competitiveness after five years of coaching by IBM consultants. In December 1998, a detailed study was made of Huawei's operations in order to get to the bottom of the problems in this area. This was prior to IBM's reform of the supply chain. It turned out that Huawei delivered products on time to only 30 percent of orders, whereas world-class enterprises maintained an average of 90 percent. Huawei turned inventory over 3.6 times per year, whereas world-class enterprises averaged 9.4 times per year. Huawei's order fulfillment period was between 20 and 25 days, whereas in world-class enterprises it was around 10 days.

In December 2003, IBM consultants provided Huawei with the data from its most recent test. Timely delivery of goods was up to 65 percent; inventory was now being turned over 5.7 times per year, order fulfillment had been shortened to 17 days. Although there was still a gap with world-class enterprises, the IBM consultants felt that Huawei had progressed at a fast pace. In their experience, it took from 7 to 10 years for an enterprise with which they had worked to transform itself from a quantity production type to an innovative-type enterprise. The extent to which Huawei had changed in just five years was rarely seen. At this speed, and another two years of "polishing," Huawei should be able to enter the ranks of world-class companies in the year 2006.

Huawei's CMM reform also brought good news. The R&D Institute in India was the first to pass the five-rank CMM international certification process. After that success, Huawei copied the same model in 17 other R&D centers including the Chinese cities of Beijing, Shenzhen, Shanghai, Nanjing, Xi'an, Chengdu, Wuhan, and in several countries around the world including the United States, Sweden, and Russia. In March 2004, after three years of hard work, all of Huawei's R&D institutes had passed the five-rank CMM international certification process.

Given these advances, the management reform that Ren initiated now entered a stage of positive reinforcement, a virtuous cycle. By 2004, Huawei also completely emerged from winter. It was greeting a bright and shining springtime. The snow was melting and everything looked fresh and new.

In 2003, Huawei's income went from negative growth to positive as the company took in USD 3.8 billion in sales. In 2004, results were even more astonishing, when sales reached USD 5.58 billion. Of this amount, 41 percent, or USD 2.28 billion, was from sales overseas.

PART THREE

GOING INTERNATIONAL

The path to becoming an international company was inevitably going to be hard and fraught with setbacks. Nevertheless, it was a necessary path if one wanted to be one of the top 500 companies in the world. Despite a host of hardships, Huawei bravely took the first step.

CHAPTER ELEVEN

CROSSING THE PACIFIC

The moment we set sail across the great ocean, our attack ran into problems. We had not yet stripped ourselves of the old "guerrilla" style management but also had not yet set up an international style of management. Our employees were not professional yet we could not recruit internationals effectively. Nevertheless, we could not wait for all these things to be perfect before launching the attack. Instead, we had to learn in the midst of the battle, become familiar with the markets and win them over, train up and create ranks of experienced troops. If we did not have "internationalized troops" within three to five years, once the China market was saturated, we would simply be sitting around waiting to die.

—REN ZHENGFEI

In 1995, when the market for program control switches in China was hot, Ren Zhengfei was already raising the alarm: "Once China's market for these switches is saturated, what do we intend to live off?" In January 1996, after the mass resignation of the marketing department, he appointed Li Li, who had formerly been head of the office in Shandong, to be overseas marketing director. He began to prepare for opening up international markets. At that time, however, Huawei was utterly at a loss as to how to penetrate international markets. All it had ever done was participate in two rather small international exhibitions, one in Beijing and the other in Geneva. In line with the strategy of taking the easy ones first, the low-hanging fruit, and then later the harder ones, Ren Zhengfei aimed first at Russia, Yugoslavia, Brazil, and Hong Kong. These became the four key markets that Huawei was intending to encircle.

From then on, Huawei set foot on the long and arduous course of going international.

Starting Out in Hong Kong

Hong Kong's telecom industry was run by an operator that was controlled by the British East Asia Telecom Bureau, in that it held a controlling block of shares. Due to historic reasons, it had held sole operating rights to Hong Kong's fixed line telephone business for seventy years. As the day for Hong Kong's return to China fast approached, Hong Kong Telecom proposed to the government in charge at the time that it take in other investors, on a compensated basis, to participate in competition. After the old entity was adequately compensated, it could end its monopoly of the land-line market in advance of the transfer of Hong Kong back to China.

In order to protect investors' rights and interests and due to various considerations including fair competition and the desire to increase fiscal revenues, Hong Kong's government at the time quickly decided that in May, 1996, it would hold a public auction for three operating licenses to the fixed-line telephone business. At that time, Li Ka-shing was in the midst of major expansion. After learning this news, he naturally felt this was an opportunity that he could not pass up. He immediately made arrangements for Hutchison Whampoa to put in a high bid in order to win the license.

For the first time since the Opium War, Hong Kong people could finally participate in this kind of business. Naturally, this was a good thing, but at the same time Hutchison also faced a severe challenge. Within three months, it had to invest USD 36 million in building a large-scale commercial network that could cover Hong Kong's central business district. This project combined the Internet, digital telecom, and consolidated access business into one. It would not get an operating license from the Hong Kong telecom management bureau if it did not meet standards both in terms of operating equipment and in terms of the timeline for building out the infrastructure. It then would not be able to get the operating license and naturally would be unable to go head to head with Hong Kong Telecom in competition.

Hutchison Telecom therefore first went to Siemens and Alcatel for assistance, but the result was disappointing. Not only was six months the minimum length of time needed, but the price they asked was not so modest. Although Hutchison had some familiarity with Huawei inside China, a company whose "horns were just beginning to emerge," it had no confidence whatsoever in Huawei's quality. As the deadline for acceptance of the bid approached, however, in between a rock and a hard place, Hutchison Telecom opted to "try" the company, given that it had no alternative.

When Huawei received this order from Hong Kong, Huawei people first spent awhile celebrating. In order to guarantee quality, the company then pulled together a group of their best and brightest engineers, all of whom had experience in setting up telecom bureaus. They called this the "team for storming the pass." The team purchased several dozen sets of pajamas and soon began spending day and night at the office. At the end of 1996, Huawei completed Hong Kong's business-network project on time.

This Hong Kong project gave Huawei people some disciplined training and it also opened the eyes of Hutchison Telecom. Nevertheless, this was not truly "going overseas to open up markets." Huawei was fully aware that it had a long road ahead of it before it was international.

Icy Russia

In April, 1996, China and Russia signed an agreement called the "Sino-Russian Joint Declaration," which established a mutual strategic cooperation partnership between China and Russia for the 21st century. Both countries now entered what was called a "new period of historical development." In May, 1996, Russia held the "Eighth Moscow International Telecom Exhibition," which naturally became Huawei's first move in opening up markets overseas.

Politics is politics while business is business, however. Just at this time, quantities of inferior and pirated goods from China were flooding

Russian markets. Some unscrupulous Chinese businessmen caused the Russian people to lose faith in China by undertaking shady deals, to the point that the Russian government made an announcement on national television encouraging the Russian people not to buy Chinese goods.

Given the commercial environment at the time, it was unavoidable that Huawei should come up against a wall in this first attempt at business in Russia.

On the eve of the opening ceremony of the exhibition, Ren Zhengfei stepped onto Russian soil for the first time in his life, but was nevertheless unable to summon much excitement. The head of Huawei's promotion department in charge of Huawei's booth at the exhibition, Zhu Jianping, had told him about the pervasive atmosphere in Moscow. Some stores turned away in disgust as soon as they heard you were Chinese; some prosperous shops even hung signs above their doors saying they did not sell any Chinese products, in order to maintain their own reputation.

"We were up against a negative tide of fake and inferior goods from China. The bad smell of China's worst gangsters[1] had preceded us just as we were trying to enter the Russian market. Who knew how long and how much work it was going to take to wash off that smell?"

After getting the report back on the exhibition from the Moscow team, Ren Zhengfei was highly disheartened. Facts were soon to corroborate his worst fears.

After the opening ceremony of the exhibition, Russia's television station, TASS News Service, Pravda, and other forms of media in Russia wrote up reports on the scene and the head of Russia's Post and Telecommunications Ministry visited Huawei's booth and even had his photograph taken shaking hands with Ren Zhengfei. Once the exhibition was over, and the head of Huawei's office in Russia, Liang Guoshi, applied for permission to install Huawei equipment in the Russian network, he met with a firm refusal. Finally, only after China's ambassador to Russia intervened on behalf of Huawei, Ambassador Li Fenglin, did Russia finally grant Huawei permission to set up a joint venture called the Bei-tuo Huawei. This was under the guise of "technology transfer" and was, moreover, with Russia's number two, not number one, switch

---

[1]The term *liu mang* broadly refers to people who operate outside the law.

manufacturer. Political considerations were the only thing that broke the stalemate, and even then not in the most ideal way.

Huawei now had "a ticket" to the game, but Russia was in fact on the brink of economic collapse. This was due to the influence of the Asian financial crisis and other considerations. The ruble was losing value dramatically. European and American telecom giants were leaving the field, but Huawei instead decided to stay behind.

After six years of hard work in the "frozen wastes," Huawei finally was able to regard Russia as one of its largest "grain-producing regions," largest producers of revenue. In the long period between 1997 and 2002, however, the company only concluded USD 38 in business, namely, a contract for batteries. That figure tells just how long and tough the Russian road turned out to be.

One Defeat after Another

Starting in 1998, Huawei sent its heavy artillery into Southeast Asia, East Asia, Africa, and Latin America in order to try to open up those markets. At the same time, it dispatched teams to Europe and the United States to see if there were any "business cracks" that it might work its way into. Although the company had done plenty of thought preparation prior to "striding out into the world," the setbacks were far beyond what they had anticipated.

This was the first time Huawei had been outside the doors of China. Overseas customers had no knowledge at all of the company, which made it hard to try and explain company products. After a period of time in which Huawei climbed the slopes and was beaten backwards, the company finally began to understand the situation. Even though Asian, African, and Latin American markets were backwards and therefore had good potential, the majority of telecom business was already monopolized by European and American multinationals. Those countries themselves did not operate their own telecom businesses.

For example, such operators as AT&T, Global Telecom, and Sprint not only covered North America but were all investing in and operating systems in more than 110 other countries and regions.

Historically, 58 of these countries and regions had been under British rule as colonies. As a result, old-name British companies not only were firmly entrenched in Europe but also in Africa, Oceania, Indonesia, Burma, Singapore, and China's Hong Kong, among others. Meanwhile, South America and Latin America contained former colonies of Spain and Portugal, so that not only were the Spanish and Portuguese telecom companies in Brazil, Argentina, and others, as core operators, but a large amount of their business extended into Mexico, Colombia, and other Latin American countries.

Since the technology behind telecommunications had been developing for over one hundred years already, the influence of brands was highly embedded in people's minds. In purchasing equipment, European operators first thought of Siemens, Ericsson, and Alcatel. American operators leaned in the direction of Lucent, Nortel, and Motorola. As far as the eye could see, Huawei people could only see territory that was already occupied and carved up by others. Despite very hard work, repeated battles and ongoing defeats, it remained very difficult to break into the international market.

One can imagine that a corporation from a developing country, and especially a corporation engaged in high-tech business, needed not only an outstanding corporate brand and product brands but also a powerful national brand behind it if it was going to compete in international markets. What's more, all three of these "brands" had to be supported in turn with corporate strength. Despite having the most outstanding international talent, the difficulty of breaking into this situation was out of the ordinary. Many companies from other developing countries had already been washed out in the course of a painful process. Huawei persevered.

Paying with Your Life

The market was already blanketed by an absolute forest of multinationals. All of these participated in monopoly markets that were heavily influenced by government relationships. If Huawei intended to survive, it clearly had to find opportunities in places that multinationals had decided to ignore, places that were simply too remote, too turbulent, or that had terrible natural environments. Choosing this path not only meant blood, sweat, and tears, but also, at times, even lives.

In those days, since almost every international flight had Huawei personnel on it, it was unavoidable that Huawei should lose people in plane crashes. At peak times, over one thousand Huawei people might be flying here and there around the globe. Some of the losses traumatized Huawei people in ways that no medicine could treat.

More than ten Huawei employees, Chinese and foreign, have been killed in plane accidents in Africa and Latin America since 1998. On March 8, 2014, when the Malaysian flight went down, killing all on board, two Huawei passengers were among those who lost their lives, Liu Sheng and Gao Jiachun. Among the plane accidents, however, the one involving Lü Xiaofeng is considered somewhat legendary in the company.

On May 7, 2002, Lü Xiaofeng was on board an Egyptian Airline flight going from Cairo to Tunis when the flight ran into a storm and went down. Fifteen passengers were killed, but Lü Xiaofeng was fortunate enough to survive. As he was escaping the wreckage, he tore off his jacket and gave it to a two-year-old Tunisian child, which was later highly publicized by African media. As soon as Ren Zhengfei heard about this incident, he made a phone call to make sure Lü was all right, and when Ren Zhengfei later visited the African region, he personally gave Lü Xiaofeng a new coat. In 2005, when Lü Xiaofeng went to Algeria to try to open up the wireless market there, he again ran into a disaster, this time an earthquake that was 5.8 on the Richter scale, with more than 400 aftershocks. European and American telecom giants hastily left the area, but Lü Xiaofeng stayed on, living in a tent and discussing ways to rebuild the telecom system. By such heroics, and by paying with people's very lives, Huawei carved out a market for itself in international markets.

In 2006, Ren Zhengfei delivered a talk at an internal corporate meeting that he called, "An R&D orientation that seeks truth from the facts, and twenty years of hard work." In this, he recalled some of the incidents that made up the difficult task of opening up overseas markets. "Starting in 1996, a large number of Huawei people began to work overseas. They left families and loved ones behind and worked in such places as Africa, still in the midst of the AIDs crisis, Iraq, where the smoke of war still lingered, Indonesia, after the tsunami, and others. Some of our employees were hit by attacks from renegades and had to be stitched up. Some were sleeping in their dormitories when scoundrels broke down the door and stole everything, guns in hand. Some were wounded in terrorist bombings. In Africa, more than 70 percent of our employees came down with malaria."

Ren Zhengfei often quoted poetry to motivate employees, stanzas that paralleled Huawei's process of "internationalizing" in that they spoke of valor and perseverance. Looking back on it now, we can see that Huawei's emergence on the international stage relied heavily on this kind of valor, this willingness to get up and fight again, this will to do what others did not dare to do and to sacrifice oneself.

Without such spirit, today's Huawei would not have been possible.

Introducing Oneself on the International Stage

Disasters did not stop Huawei's forward advance. What did keep the company from progressing for awhile was a very considerable cultural gap. Part of this related to the outside world's impression of China, a country that had been in turmoil since the Opium War and therefore was seen as impoverished and backwards. Another part related to the negative feelings people had about "Made in China" due to the shoddy goods that China threw onto international markets in massive quantities in the mid-1990s.

The Huawei personnel who were responsible for opening up the European market in those years look back on the challenges now. The most difficult part, they say, was not isolation as they lived overseas but just that they could not get in to see customers. Sometimes, when they finally did get an opportunity to meet with technical personnel face to face, they found that the language barrier kept them from communicating effectively. This only increased customers' doubts about the ability of a Chinese company to produce high-tech products made with its own intellectual property.

Since the policy of "striding forth into the world" was being held back in this way, Huawei had to find ways to bridge the cultural gap. One of the main ways to turn around the firmly entrenched impression that people had of Chinese corporations, and evaluate Huawei's products in a more objective and fair way, was to participate in international telecom exhibitions. This "stuck a pole in the ground so you could see the actual shadow." Starting in 1998, Huawei began participating in every exhibition put on around the world. Participation became an iron law at the company that has persisted to this day.

Generally speaking, the presentation of the major European and American telecom giants at such exhibitions is unostentatious, given that these companies already have a massive advantage in branding and cutting-edge technology. They have smaller booths and fewer people. Huawei's approach has been different. In order to emphasize its real power, Huawei has consistently had far larger booths than others and also exhibits its most recent technologies. In order to strengthen the impression that customers take home, it rents special areas within the highly expensive exhibition hall to do on-site demonstrations. If it can attract the thoughtful attention of customers, Huawei has not worried about costs, which have been fairly massive. In 2000, Huawei invested some RMB 100 million in a Hong Kong international telecom exhibition, yet the company's overseas sales revenue in that year came to a paltry RMB 400 million.

Naturally, such large investment was bound to draw attention. As one Huawei employee put it after a particular international telecom exhibition, "Many overseas customers fundamentally did not understand Huawei. Once they had seen our exhibition, they had a strong initial impression. Later, when we had the chance to introduce our technology and products to them, that initial impression made it a lot easier for them to take in what we were saying."

In addition to participating in exhibitions, Huawei also hosted a number of international forums and industry seminars. These were an important measure in strengthening customers' acceptance of Huawei's products and technologies. For example, in 2000, after the IT bubble burst, Huawei grabbed the opportunity and invited more than forty customers from twenty-five countries to participate in an international forum in (Mangu and Wenlai), as well as a "Wenlai NGN international forum." By inviting customers to come on-site to analyze Huawei's successful experiences, Huawei was able to begin to bridge the communication gap with international customers.

In order to expand the influence of its brand, Huawei revised its long-standing policy of keeping a low profile within a given country. It began to place ads in such international media as the *Washington Post*, the *Economist*, the main newspaper in Thailand, and others. It carried out a series of activities aimed at promoting its image. At the same time, it participated in international standards organizations of all kinds. It has actively promoted the development of standards in the telecom industry in order to improve its own "right to speak" in this community. Since 2001, Huawei has been very active in the ITU (International Telecom Union), 3GPP (Third Generation Cooperative Partner Plan), IEEE (Institute of Electrical and Electronics Engineers), the IETF (Internet Engineering Task Force,) ETSI (European Telecom Standards Institute), OMA (Open Mobile Alliance), and so on. It has put forth 28,000 suggestions to these international standards organizations and made outstanding contributions to the formulation of 3G, 4G, and NGN (Next-generation Network) standards. Based on this extensive activity, Huawei has been elected chair, vice-chair, or a member of the board of many of the 148 international standards organizations. The company has played an increasingly important role in the formulation of global standards.

Opening Up a New Silk Road

Bridging the cultural gap and playing an active role on the international stage began to give Huawei a measure of name recognition. At the same time, the company began to recognize that it would be impossible to become a well known international brand by relying solely on its own efforts. It would be necessary to link the fate of the company closely to the fate of the country itself.

To do that, in 2001, Huawei began a promotion effort that it called Opening Up a New Silk Road. This involved inviting customers to come to China and take actual tours to see China's major accomplishments since the start of reform and opening up in such cities as Hong Kong, Beijing, Shanghai, Xi'an, and Dalian. People were invited to see the earth-shaking changes in the country for themselves. At each place, Huawei would organize visits to the offices of three different companies, China Telecom, China Mobile, and China Unicom. The aim was to give them a visceral understanding of how Huawei equipment was being used on site to service the telecom needs of 1.3 billion people. It was to make them realize that Huawei's equipment was fully at the level of world standards. The final stop on such a trip was a visit to Huawei's headquarters in Shenzhen.

When customers entered Huawei's Bantian production site, which cost RMB 10 billion and which covered 1.3 square kilometers, they were impressed. It had first-class production lines in highly modern factories, and an exhibition hall with a helicopter landing on top. People became more willing to engage in cooperation with Huawei.

Hardware facilities impressed customers, but Huawei also did not beat around the bush when it came to "the software side of things." Huawei set up a prayer hall for Islamic guests in its reception area, for example. In 1999, a senior business official from Saudi Arabia visited Huawei. He noted to the staff accompanying him, "If a country intends to become prosperous, it has to have a fighting spirit. This is not something that is done in one generation, though. It has to be pursued generation after generation." This was in appreciation of the cots that R&D people kept under their desks, but also the comprehensive approach of the company's efforts.

In 2001, the Minister of Russia's Ministry of Posts and Telecommunications visited Huawei. Prior to his arrival, he was aware of the cost advantage of Huawei's products due to the report that his staff had prepared for him. He was still doubtful about the company's technology and its ability to meet the standards of such giants as Ericsson and Siemens. He made a personal and detailed inspection of Huawei's R&D Center. Surprised by what he saw there, he signed a contract for GSM technology with Huawei for more than USD 10 million on the spot. Huawei's sales began to rebound. In the first half of 2002 alone, the company sold more than USD 300 million in wireless products.

Some customers were unable to participate in the Silk Road tours for one reason or another. To enable them to see Huawei products in person, in 2001 the company launched another initiative that it called the New Orient Express. This equipped the cars of a specially commissioned train with Huawei's latest equipment and it then travelled throughout Europe, making stops in major cities. At each stop, sales staff would deliver detailed presentations on Huawei's new technologies and its price-to-functionality ratios. Once customers expressed interest, sales staff would invite them to participate in a Silk Road tour at the appropriate time. When this second stage of the process was over, customers generally were ready to sign up with Huawei for impressive amounts of orders.

Overseas Markets Do Not Accept Opportunism

This strategy of both "striding out into the world" and "inviting people in to China" had certain success in opening up overseas markets for Huawei. However, Ren Zhengfei was highly aware that Huawei still was far from being able to share one-third of the world's market together with European and American giants in the business. The path to success was full of uncertainties, but one unalterable fact remained, namely that overseas markets rejected any form of opportunism.

Ren Zhengfei therefore declared to his staff that they were fighting a prolonged war of attrition. It had to be fought on the basis of ground already firmly held by the company. This war could not be won on the basis of simply hauling in orders wherever possible, and simply going away if orders were not available. If the company was stable and consistent in making sure its quality was superior, its prices were low, and its service was excellent, sooner or later international markets would open their doors. This approach of being consistent and predictable then became the general compass for all activity in Huawei as it carried on international sales activity and engaged in multinational competition.

In 1999 and 2000, customers needed a high rate of return on investment after the Asian financial crisis. Otherwise, they were psychologically unprepared to place orders. Huawei took full advantage of this by setting its prices at 30 percent below its competition. One after another, it took over the GSM markets in Vietnam, Laos, Cambodia, and Thailand. After that, it used the same method to expand its advantage in Middle Eastern and African markets.

At the same time, Huawei remained very aware that its forces were thin when it came to tackling the European market, where "giants stood in serried ranks." It did not, therefore, engage in frontal attacks on

Audophone, British Telecom, German Telecom, or France's Telecom. Instead, it used the strategy of "surrounding the cities by taking the countryside," which in this case menat first going for the easy ones and later the hard ones. It focused its attack on mid-range operators by using flexible terms, operators who were just getting started and who were extremely attentive to price-functionality ratios. The company quickly stirred up waves in this mid-range part of the market.

Opening Europe's Gates by Using Price as a Weapon
In 1996, the U.S. government announced plans to reform the structure of its telecom business, and led the way in opening up its telecom operating markets. Various countries in Europe soon followed suit. Internet technology was changing rapidly at this time, providing a massive shock to traditional telecom markets. Some old-name brands were soon in trouble due to carrying a heavy debt burden, which provided the opportunity for younger and more nimble new-style companies to move into their space. Neuf was one such company.

Neuf supplied telecom operators with optical-fiber products as a wholesaler and installer in its earliest period. As time went on, cable television, Internet, and traditional telephone voice services began to be integrated into one system. Neuf then decided to shift into this new arena. Its first project was to build a backbone transmission network that covered all of France. To do this, the company invited various "giants" to participate in a bid, including Alcatel, Lucent, and Ericsson. The price they came up with far exceeded Neuf's budget, however. Encouraged by a core member of the board, Neuf then asked Huawei for a bid, and Huawei submitted a bid that could not be matched.

Neuf was extremely interested, but it also was still doubtful about Huawei's capabilities at networking technology. This particular project would be a matter of life or death to Neuf, since it would make or break the company's future reputation. Any stoppage in optical transmission would deal a deathblow to Neuf since it was just in the midst of changing its corporate direction. To dispel these doubts, Huawei proposed to set up test networks in the two cities of Lyons and Marseilles. After initiating the system, it would allow the company to operate it for three months for free before final approval. This was highly persuasive to Neuf. What the company had not expected, however, was that Huawei would complete the installation and initiation of operations within two months in outstanding fashion. The operating results were quite satisfactory and the business was approved.

After this smooth success, in March, 2003, Huawei put in a bid for a backbone transmission network that could cover all of France. In this bid, which was for 4.22 million Euros, Huawei truly "emerged from its shell." It beat out such formidable adversaries as Lucent, Ericsson, and France's own Alcatel by using its clear advantage in superior quality, low price, and excellent service. Neuf's profitability and operating strength were greatly enhanced by this project, and its own business began to soar. After it successfully purchased the transmission and exchange networks of Italian and German telecom companies in France, it became the second largest operator in France, right after France Telecom in size. As Neuf's primary strategic partner and core supplier, Huawei was launched as well.[2]

Portugal's market was another success in these days. As a coastal country, Portugal is famed for its holiday vacations and receives some 20 million visitors every year, well-heeled people who come to play golf from all over the world. INQUAM was a mid-sized multinational mobile operator with headquarters in England. Although it had been fairly early in setting up GSM networks in England, Romania, France, Portugal, and Morocco, it was limited in the amount of investment it could spend on expansion and therefore was not highly competitive in the mobile market. To find new growth points, it began to explore Portugal quite actively once it learned that the country's many visitors were unhappy about not being able to access CDMA when they were there. It took the initiative

---

[2]According to Internet sources, Neuf Cegetel was legally established in 2005 following the completion of the merger between Neuf Telecom (formerly known as LDCOM) and Cegetel. As of June 2008, the company became a wholly owned subsidiary of SFR, and the brand disappeared commercially.

 LDCOM (the future Neuf Telecom) was established by the Louis Dreyfus Group in 1998, at the time of telecommunications deregulation.

 Between 2000 and 2003, French telecom went through a major consolidation. LDCOM acquired several alternate operators cheaply, which the author refers to in this section.

 Siris was France's third largest operator, which Neuf acquired from Deutsche Telekom (May 2003). Neuf purchased 9 Telecom from Telecom Italia (August 2002).

in issuing a request for tenders to the three main giants, including Nortel, but the results were disappointing since prices were far over its budget. After considerable internal discussion, the company decided to give up the idea of investing in a CDMA network.

Qualcomm, which had the exclusive patent on CDMA technology, then heard this news. It quickly came looking for INQUAM's policy team and firmly recommended Huawei to INQUAM. Moreover, it promised that it would give INQUAM preferential business terms. This made INQUAM's senior policy-making team decide to give Huawei a try. However, the team was not completely convinced about Huawei's equipment and whether or not the company could meet operational plans and scenarios for future development.

As it happened, the product-solutions proposal that Huawei put forth was a surprise. First, its overall price was one-third that of telecom giants. Second, its CDMA equipment could in fact provide coverage for all of Portugal and could also provide improved end-to-end solutions by simultaneous realization of international roaming, multimedia channels, DO, PTT, wireless telephone, and so on. In a specially prepared laboratory, INQUAM carried out dozens of tests of Huawei equipment. The consensus of the team was that results were "excellent."

The following steps were simple. INQUAM set up Europe's first CDMA business network in Europe, in Portugal. It became profitable within the anticipated time period and soon paid off the investment. As hoped, Huawei was able to set up a "model experience" within Europe. It then took advantage of this success to open up CDMA markets in other countries, including in Asia, Africa, Latin America, and North America. Meanwhile, it worked together with local established agents in Europe, and successfully opened the door to markets in Germany, Italy, Spain, and England. Its reputation in Europe rose.

Running Faster than a Lion

As described above, Huawei used price as its primary weapon to pry open a crack in certain markets and then took advantage of the opening to expand business. This was undoubtedly the best approach for small and low-end customers. When it came to high-end operators, however, price was not the primary consideration. Actual products and the services that an equipment supplier could provide were more critical for them. The issue was whether or not the supplier could help them compete with their own competitors and make them more profitable. Price then factored in as well.

AIS was a telecom operator in Thailand. Before Yingla Shinawa became Premier of Thailand in 2011, she had been CEO of this company for a number of years. In 1999, she decided to take the company forces into the mobile market and at that time her main competitor was DTAC, which had already taken over more than 45 percent of Thailand's mobile market. Clearly, as a follower in this market, not only did AIS need an extremely high price-to-functionality ratio, but it needed to have technology proposals that differentiated itself from its competition.

Huawei spent a long time researching the Thai tourism market. It discovered that there was a large market in the prepay business. As a result, it made a proposal to AIS to get into this market. Yingla's family had originally migrated from Guangdong province in China. However, she had spent many years as a student in Europe and the United States, so her inclination was to lean in the direction of Ericsson, Nokia, and Motorola when it came to brands and technology. However, she also saw that the response time of these huge companies was slow, and that their product-solutions proposals were not fully satisfactory. Given the time pressure, she finally decided to go with Huawei.

Unexpectedly, Huawei not only completed the installation of equipment within two months, but was also half a year ahead of DTAC in launching a prepaid business and thereby attracting large numbers of customers. Yingla's company successively launched other kinds of business that allowed it to increase its customer base by more than 10,000 people every single day. Within less than five months, it had recouped its investment and created a kind of miracle in the history of Thailand's telecommunications operators. As a result of this, AIS replaced DTAC to become Thailand's top mobile operator. Its market capitalization rose swiftly on the Thai stock market to become the premier share traded on the market.

Yingla strongly endorsed Huawei at an international forum held in Rangoon in 2004. "Huawei's swift reaction time, its highly effective business development, and its engineering services helped us improve our own core competitiveness." Representatives of mobile operators from twenty-five countries were assembled at this forum, and Yingla was supportive as each began to carry on business discussions with Huawei.

A similar situation soon developed in Europe, with Holland serving as the gateway in this instance. Although Holland's population was just 16 million people, its mobile market was large enough to have attracted fierce competition among European giants including Vodafone, German Telecom, France Telecom, and Holland's own Royal Telecom.

Telfort was at the time a fairly small telecom operating company with around 2.5 million mobile customers. Given the high density of Holland's population, and the extremely expensive cost of labor in the country, the question for Telfort became how to differentiate itself in competition with the crowd of giants, how to emerge as a contender in the 3G age. Given the intense competition, Ericsson was more focused on the other four larger high-end operators in terms of Holland's 3G network deployment. It naturally could not provide a product-solutions proposal to Telfort that was competitive.

In June, 2004, via introduction from a friend, Huawei's chief representative in Holland paid a call on senior management at Telfort. Once Huawei had understood Telfort's predicament in its plan to build out a network, Huawei went to work. Within one week, the company prepared extremely competitive solutions proposals for the company that were based on a distributed base of stations. These wall-hung types of distributed stations were no larger than a DVD. Installing them was light and easy. They saved on space, lowered installation costs, and were less polluting than alternatives.

This incredibly fast reaction time on the part of Huawei impressed Telfort's senior management. Teams not only visited the company but also carried out on-site inspection of Huawei's various partners in 3G commercial-use networks including SUNDAY in Hong Kong, Emtel in Mauritius, and TM in Malaysia. They expressed strong satisfaction and began to place orders. To date, their 3G orders with Huawei have exceeded 200 million Euros. Meanwhile, Huawei was able to get this business within the space of four months and successfully opened up Holland's 3G commercial-use network.

"We were correct in recognizing Huawei's sincerity and its creativity," the Chief Financial Officer of Telfort, Fandeweier later noted. "Everyone thinks of Huawei as being a company that relies on low price to win business. In fact, this is not the case. The key for us was recognizing that Huawei was able to meet its contractual obligations and to do so quickly. As for price, I can say in all honesty that Huawei's price was not in fact the lowest."

Another major opportunity then held out its hand to Huawei. On June 29, 2005, the "elder brother" among Holland's telecom operators announced that it was spending 1.12 billion Euros to purchase Telfort. Since Huawei had beaten out Alcatel and Lucent, two leaders in the European optical transmission arena, it had also won a very large order from KPN for a backbone network covering all of Holland's major cities. The reason was simple: Huawei's excellent levels of supply-chain management and its fast reaction time had played a decisive role. When KPN presented its requirements to Alcatel, Huawei, and Lucent at the same time, it asked all three to send equipment to its laboratory for testing. Huawei sent its equipment from China, yet still beat out the other two by one week, even though the others were based on the European mainland. In the end, Huawei won the order for its excellent functionality-to-price ratio.

All of this signified that not only was Huawei winning a seat at the table in Europe's fierce competition for the 3G market and digital telecom market, but its fast reactions and excellent supply-chain management was making Europe's top-tier operators begin to take a second look at this company from China.

"If a mountain goat intends to keep from being eaten by a lion, it has to run faster than the lion. If the lion intends to have a full stomach, it has to run faster than the mountain goat. If both run at the same speed, both are going to die from exhaustion." Ren Zhengfei often quoted this adage about the strong eating the weak, and survival of the fittest. It was fully born out by Huawei's experience in fighting for overseas markets.

The International Deployment of Huawei's R&D

If you do not want to be rolled by a wave, you must provide your customers with an outstanding price-to-functionality ratio. In addition, you must provide them with first-class technology and first-class product-solution proposals. Huawei people gradually came to this conclusion as the company fought numerous battles and went from constant defeats to a few successes. Due to a more deep-seated understanding of the nature of the industry, starting in 1999, Huawei therefore began to put massive investment into two simultaneous processes, internationalization of the corporation and the establishment of R&D facilities overseas.

The United States stands in the premier position when it comes to software development, while India is number two. In order to move into these locations as well, Huawei has put USD 400 million into international software-development facilities since 1999. The company set up its first R&D institute in India, located in the third, fourth, and fifth stories of the seven-star Leela Hotel in Bangalore. Between 1999 and 2012, more than 6,000 outstanding Indian software developers have worked here, in a lovely work environment that includes ancient architectural monuments and tropical foliage. Not only have these people provided enormous human resources for Huawei in terms of software development, but they have contributed to the company's business development and the upgrading of its products.

Starting in 2001, Huawei picked up the pace of international deployment of R&D. The United States was the source of CDMA, digital transmission, and cloud computing, so Huawei first set up two research centers in Silicon Valley and Dallas. Europe was the source of 3G and Ericsson was leader in this technology, so Huawei set up an R&D Institute in Stockholm. Russia was a world leader in the field of wireless frequencies, so Huawei set up an R&D facility focusing on this area in Moscow.

Although these places could not compare with India in terms of numbers of personnel and size of investment, the role these institutes played was critical in placing Huawei's R&D at the center of global new technologies. They enabled Huawei to have a finger on the pulse of new technologies, but they also allowed it to bring international talent into the company, to purchase patented technologies, and to provide technological support that was close to global customers.

The research facility in the United States is a good example. Dr. Bai Lusheng, a Chinese American, first graduated from Harvard with a PhD in 1987 and then Stanford in 1990 with a PhD in optical fiber transmission. After earning this double PhD, he joined the team at Huawei in 2001. He worked together with the R&D team and in 2002 Huawei launched optical transmission products. Their functions not only allowed optical transmission to go furthest, but longest across multiple transmissions. Moreover, they could tolerate all kinds of low-end optical fiber so that they had low cost for high functionality, and this brought massive commercial benefit to customers. It enabled Huawei to get backbone transmission orders that covered all of France, then Holland, then England. In 2008, under the personal supervision of Dr. Bai, Huawei also successfully developed the technologies called 40G and eDQPSK. Since these were solely held by Huawei, they were a kind of "shot heard round the world" as soon as they were launched. Their market penetration swiftly rose to 44 percent in terms of total international markets.

By today, Huawei has set up seventeen R&D Institutes and twenty-six Innovation Centers around the world. Huawei's strategic R&D initiative is gaining ever-more authority as the company becomes an international entity. The research orientation of each facility and the technology each focuses on is different, but they all allow Huawei to have a head start on following up new technologies and understanding the latest customer requirements. The company then goes through an open-style IPD R&D platform to carry out simultaneous work on development and coordinated strategies for market development. It thereby is able to turn commercial opportunities into commercial results at an astonishing speed.

Having the Younger Student Guide the Older Student in Conquering the World

Each of Ren Zhengfei's major initiatives starts with a modest step. The first batch of Huawei pioneers to "open up" international markets were a bit like grade-school students. They were curious about everything and they hoped to adapt to their environment as quickly as possible. On arrival in any given country, they would record all kinds of things in their personal journals—what people wore, how they communicated with one another, what their daily habits were. This mode of learning was not going to enable them to penetrate European circles very quickly, however.

Although Europe is fairly open in overall terms, each country maintains its own customs, etiquette, and habits. In Germany, for example, sales staff would appear rigid and hide-bound if they put on a suit and tie when visiting customers. The opposite was true in England. If one dressed casually there, one would be seen as not taking the business seriously, and not having the proper regard for proprieties.

After running up against a wall several times, Huawei people began to realize that the fastest and most effective way to get on good terms with Europeans and develop business would be to hire Europeans. This became patently obvious. Recruiting people who were already familiar with local customs and knew the local situation was best. Moreover, such people generally were "high quality" and very professional. They could train up Huawei people and play multiple roles as interpreters and guides at the same time.

One person in Huawei's Frankfurt office named Suolun Pierxier was just such a person. He had been a product manager for a telecom group in Europe before coming to Huawei. After entering Huawei in 2002, he not only introduced Huawei people with great ease to German customers and initiated technical exchanges, but he also lifted the caliber of Huawei's bids for German business, and helped daily in upgrading and standardizing its capabilities. "Not only is he worthy of our trust as a colleague and as a friend, but his conduct is fully in line with his reputation." This endorsement by a member of Huawei's Frankfurt representative office was spoken from the heart.

In addition to recruiting a large number of foreign nationals in any given country, another of Huawei's strategic moves was to hire Chinese overseas students who were studying in that location. This strategy was actually given a name: "Having the younger student guide the older student in conquering the world."

The head of Huawei's representative office in Holland, Chen Haijun, was formerly such an overseas student, studying in Europe.

After graduation, he worked several years for the Holland division of Ericsson. The pay was excellent and living conditions were comfortable. Nevertheless, these material considerations were not enough to overcome the feeling that he was in another country, apart from his own. When he learned that Huawei would be hiring in Holland, he promptly sent over his resume. Once he joined the company, Huawei's presence in Holland expanded quickly. For example, Huawei signed the order with Telfort that is described above for over 200 million Euros for 3G equipment. Huawei's quick reactions and price-to-functionality ratio were important in getting this order, but it also had a lot to do with Chen Haijun's close familiarity with the local scene.

"Having the younger guide the elder." This approach, encapsulated in Ren Zhengfei's pithy phrase, was instrumental in paving the way for Huawei's frontline teams. It also enabled Huawei to move into a development stage of "scaling up" on a solid basis.

Starting in January, 2001, Huawei announced a series of incentives aimed at encouraging staff to move abroad. These appealed to people to be "brave soldiers," since "top generals need brave soldiers under them." In addition to compensation packages that were tailored to the local economic conditions in each different country, Huawei offered subsidies of between USD 50 and 200 per day. Huawei's rules now specified that direct relatives of Huawei employees could go abroad to visit them three times a year. The cost of their flights would be covered. At the same time, any Huawei employee who had worked abroad for three years could receive a "family resettlement fee" of RMB 150,000. These policies were a tremendous incentive in getting Huawei people to move overseas to work.

Meanwhile, Huawei adhered to two principles when it promoted people and assigned people to jobs. First, it evaluated whether or not the person had taken the initiative himself in applying to move overseas. Second, it ruled that cadres who had not worked abroad were not qualified for promotions. A catchy rhyming couplet therefore went around within the company to the effect that if you wanted to move up the ladder, you had better move abroad right away.

In order to have senior cadres set the example, Ren Zhengfei first sent a batch of senior managers overseas. Then, one by one, he began to send the heads of domestic offices overseas to various parts of the world.

Those office managers who had not yet been assigned overseas posts found their post office boxes constantly stuffed with notifications from the marketing department, telling them to get ready to leave at any time.

In order to enable the "younger students" (Huawei employees) to guide their "elders" (overseas Chinese) in a more effective way in opening up international markets, Huawei launched a rigorous program of English training. Both English and Chinese were now used in the IT system, on the websites, and in internal documents, but in addition, passable English also became a "hard constraint" in determining a person's performance evaluation and his ability to go overseas. Essentially all Huawei people, including the drivers who met foreign visitors at the airport, were now required to take exams in conversational English in order to be promoted.

A fairly professional cadre of people was beginning to form within Huawei. In all areas—R&D, technology, marketing and sales—people began to combine Chinese and western elements in how they operated. This seasoned force, "not only able to march but also able to fight," was no longer just a dream for Ren Zhengfei. From a mere spark, it was beginning to set international markets on fire.

CHAPTER TWELVE

LANDING FORCES ON NORTH AMERICA, BUT ENCOUNTERING POWERFUL RESISTANCE

We lack the basic R&D capacity of Lucent. Our products may be advanced now, but this is temporary. If we do not occupy certain markets as fast as possible by taking advantage of our lead, and if we do not increase investment to consolidate our position, every little advantage we enjoy will be whittled away. If we do not work hard, we will be struck a mortal blow. We therefore must attack when it is time to attack. All of our outstanding sons and daughters must bravely fight to gain orders, without any thought of surrendering turf.

—REN ZHENGFEI

After five years of "battling upwards and being cast downwards," Huawei's overseas sales figures finally reached USD 244 million in 2001. Although most of these sales came from non-mainstream markets in Asia, Africa, Latin America, and Europe, they were still orders. They gave Ren Zhengfei a glimmer of hope in the midst of a telecom market that was universally in the doldrums.

In the United States, telecom had undergone ten years of abnormally fast growth, too much investment, and overproduction. As a result, when the IT bubble burst, not a single one of the European and American telecom giants avoided the damage. Operators reduced their spending. All now tried to do more with less money. The monopoly-type model that had previously applied to western companies, one of "high investment and high returns," was now hard to continue. Instead, a business model that relied more on "outstanding quality, low prices, and excellent service" was winning the day, that is, the strategy espoused by Ren Zhengfei.

One thing that particularly contributed to his confidence was that the traditional telecom markets were already becoming saturated, while Internet markets were instead "greeting the springtime of broadband." After the September 11 incident, the Bank of America, Citibank, and Chase Manhattan began setting up online systems on a global basis in order to improve their figures. In 2001 alone, these three banks procured USD 1 billion worth of digital telecom products. Because of this, using router products to open up the huge American market was undoubtedly a rare opportunity for Huawei. However, this most fertile "grain-producing region in the world" already had somebody closely guarding it, specifically, the global network giant Cisco.

The Only Thing Different Is Price

In 1999, Huawei was attacking the digital telecom markets in the Asia Pacific region, Africa, and Europe. As this was going on, Ren Zhengfei sent one of the old timers from Huawei's marketing department, Chang Zheng, into the American market. In January, 2001, Ren Zhengfei made a speech at the event to "Welcome back and send off valorous soldiers overseas," at which he also announced he was sending one of Huawei's most "wolf like" generals to America, namely Xu Zhan. A wolf team was soon assembled, and Huawei swiftly began to open up the North American market.

In addition to this, Huawei began to put ads in media publications that also carried Cisco ads. Huawei's tag line was, "The only thing different is price."

In 2001, Huawei's overseas sales were USD 244 million. These rose dramatically to USD 552 million in 2002. Meanwhile, in 2002, Cisco's market penetration dropped from 80 percent in the first quarter to 73 percent in the second quarter. Faced with the aggressive force of

Huawei, Cisco's board of directors undertook considerable internal discussion and came up with a solution.

At a telecom exhibition in Atlanta, a person named Qian Bosi reviewed Huawei's booth. He asked the person on duty there about Huawei's high-end, middle, and low-end router technology and then hurriedly left. Not long after, Cisco formulated a series of measures to deal with Huawei's strategic attack. Soon after that, a number of websites in the industry revealed that Cisco's routers were similar to Huawei's in appearance, coding, and functionality. Cisco also formally made a public announcement on its website: Huawei's routers had used core technology from Cisco.

On December 10, 2002, several senior managers from Cisco had an emergency meeting with Ren Zhengfei, Guo Ping, and other senior Huawei people at the Shangrila Hotel in Shenzhen. Cisco's team formally declared that Huawei had infringed Cisco's intellectual property rights. Moreover, they presented certain demands, including that Huawei "admit the infringement, that it pay compensation, and that it stop selling these products." Faced with this hard-line attitude on the part of Cisco, Ren Zhengfei expressed his own opinion in quite candid terms.

"Just as Huawei protects its own intellectual property rights, it also respects the intellectual property rights of others. Huawei is very willing to resolve this dispute between the two sides in accordance with the facts, and it welcomes Cisco to conduct an inspection of our products."

After this meeting, as negotiations were going forward, Huawei was proactive in examining its own design documentation while at the same time it voluntarily called back some of the disputed routers from the United States. However, it firmly denied that there had been any infringement. After several fierce debates ended without results, the two sides parted with unhappy feelings.

Cisco's Patent Suit

On January 23, 2003, Cisco formally filed a suit against Huawei with the Marshall Division of the U.S. District Court for the Eastern District of Texas. This day was traditionally the last day of the "small year" in the old Chinese calendar.[1] In a 77-page document, Cisco accused Huawei of infringing its intellectual property rights. It listed twenty-one charges relating to copyright, patents, and trademarks. It also made compensation demands in the material it submitted to the court: Huawei should pay a penalty for infringing Cisco's patents; Huawei should pay Cisco compensatory damages for such infringement; Huawei should pay in full for the economic loss caused to Cisco; Huawei should pay reasonable legal fees associated with this case; Huawei should assist Cisco in recovery of damage funds and other damage-type compensation...

After the head of Huawei's legal department for international affairs examined the documents in detail, he concluded, "If you added up all of these demands, the compensation would come to an astronomical figure."

Since the spring festival holiday was underway in China, all department heads had already left for their year-end holidays. Ren Zhengfei and Sun Yafang were overseas on an investigation trip. Once they heard this news, however, they hurried back to China. Several senior cadres who lived in distant places now also caught flights and all quickly converged on Shenzhen.

Everybody was quite clear on the matter. If Huawei lost this suit, not only would the company be expelled from the U.S. market but its reputation would be destroyed and the implications would not stop at international markets. Moreover, Huawei was just at a critical position in that cash flow was extremely tight. If it lost the case, the massive amount of compensation would destabilize its martial spirit. The impact would spread to the price at which shares could be converted, and Huawei's bankruptcy would only be a matter of time. This multinational suit was therefore not just a serious blow but potentially a mortal blow to the company.

After several days of in-depth discussions, Ren Zhengfei decided that this was a battle the company could not afford to dodge. If Huawei was to achieve victory, it could not retreat out of weakness. Instead, it must be brave enough to "meet force with force." He therefore assigned Guo Ping to the job of responding to the suit, and he organized senior people

---

[1]This refers to the last day of the lunar calendar in which the year has only 29 days, that is, an "off year."

from several departments to form a team to deal with the suit. These were from the intellectual property department, the legal affairs department, the public relations department, and several people from R&D and the international sales department.

Before this team left for the United States, Ren Zhengfei advised Guo Ping as follows: "You will have to bear the pain and the ridicule of this event. When you come back, I want you to be smiling. You must let the facts speak for themselves!" This elder in the company had fought alongside Ren Zhengfei from north to south, and weathered all kinds of battles. He had himself done thorough thought preparation. Nevertheless, once he arrived with his delegation in Texas, he found that the matter had progressed far beyond what he could have imagined.

By this time, all of the major newspaper and publications in the United States were reporting in detail on the "infringement by Huawei." Cisco's spokesman was also making statements against Huawei in the local media, on websites, and throughout more than 100 branch organizations of global media. These were broadcasting news of Huawei's infringement. "Cisco does not file lawsuits without reason. Huawei has unlawfully copied Cisco's intellectual property and refused Cisco's numerous attempts to resolve these issues. As a result, Cisco has no choice but to protect its technology and the interests of its shareholders through legal action."

Chinese "Gongfu" versus American "Boxing"

This clearly was now a case of Chinese martial arts going up against American-style boxing. It was a highly provocative test of strength.

In targeting Cisco's powerful publicity attack, Huawei emerged quite vigorously from its traditionally closed nature. It began to appear frequently in China's various forms of media. By standing up for itself, it sought to save its name within China. Senior Huawei officials met with representatives of China's major media in one-on-one interviews. The magazine *IT Manager's World* published a cover story written by the veteran journalist Sun Li. This was called "Huawei, and Cisco engage in a commercial bout, using the pretext of intellectual property rights."

The intellectual property rights expert, Dr. Wang Xianlin, published an article called "Reflections on monopoly issues due to private agreements in the dispute between Cisco and Huawei." The investigative department of *Telecom World* and the special reports department of *Investment and Cooperation* also published a White Paper that opposed such private agreements on February 13, 2003. This examined the impact of private agreements on the development of the industry with a focus on market competition and technology standards. It looked at national security issues, as well as Cisco's intent to use the "weapon of patents" to impose a monopoly strategy on Huawei. These articles and studies played a critical role in guiding the domestic media to look at the real issues head on as they investigated this case. Soon after, Huawei released a number of documents to domestic media so that they could pass on the information to the public. These included a description of factors that were antecedent to the suit, the developments to date, and relevant internal pieces of information. These helped support Huawei and improve its image within Chinese domestic media, by using the information so derived to put a "true face on the matter."

At this same time, Huawei also invited one of the most experienced law firms in intellectual property disputes to come to Shenzhen, namely Heller Ehrman. A team from this law firm visited Huawei's production base and R&D facilities. It learned how Huawei had built a first-class technology and R&D capability by going through IPD and CMM. After undertaking a thorough and conscientious investigation of Huawei's overall strength, and weighing all factors, this international law firm was not only astonished but it accepted Huawei's appointment. The top attorney in the field of intellectual property rights then proposed the idea of using a counter-attack on the basis of "private agreements." This would take aim at Cisco's monopoly behavior as revealed by such private agreements. The strategy was in fact able to achieve a fundamental turnaround in the law suit.

At the same time, Huawei invited a third-party expert to come to Huawei, the digital telecom professor Dennis Ailisen from Stanford University. He did a comparative analysis of Cisco's IOS and Huawei's VRP platforms and the new and old versions. His conclusion:

Huawei's VRP platform had 2 million lines of original code, while Cisco's IOS used 20 million lines. This evidence was a considerable surprise to both the judge and the legal teams once Dennis Ailisen produced it in court.

On March 20, 2003, Huawei and 3Com announced that they were setting up a joint venture company. Together, they would be researching digital telecom products, as well as carrying out production, sales, and marketing on a global basis. On March 24, the CEO of 3Com, Bruce Claflin, appeared in court as a witness. "I have been to Huawei countless times and I have certified technology there for eight months. In my several decades of professional experience, I can ensure you that Huawei is trustworthy. If there were any existing infringement of rights, 3Com would not be taking this massive risk and setting up a joint venture with the company."

By now, the legal authorities in the case had all begun to regard Huawei with a softer look in the eye.

The Price Paid for Victory

Cisco's board of directors had been highly confident of victory prior to launching the company's forces against Huawei. The manner in which Huawei responded took them by surprise, however. Huawei used not only adept Chinese *gong-fu* but also sharp American pugilist techniques.

Cisco's intent had been to force Huawei to lose the suit quickly in the United States, on American ground. This was now complicated by the way in which Huawei set up a joint venture with 3Com. The joint venture signified that even were Huawei to lose the suit it would still be able to continue to expand in the United States under the name of its joint venture. What's more, Huawei possessed its own core technology, which it had developed itself, and Cisco did not in fact have in hand irrefutable facts that showed infringement by Huawei.

This series of unforeseen events made Cisco aware that Huawei was definitely not dead in the water. What's more, the company was proving adroit at hatching schemes and was clearly going to be difficult to deal with. If the suit was not quickly concluded, this would lead to losses on both sides. Given these considerations, even though each inch of ground still had to be contested in court, both sides already knew in their hearts that conciliatory efforts outside of court would be necessary.

After two initial hearings of evidence, the case formally began on June 7, 2003. The court denied Cisco's request for it to issue an order restraining Huawei from selling products, among other things. However, with regard to appraising evidence on the disputed source code, the final determination on this would decide who was going to win and who would lose. In order to arrive at such a determination, both Cisco and Huawei signed an agreement temporarily halting the suit, pending more information. Under the supervision of lawyers on both sides, a third-party expert carried out a stringent comparison of source code. In the end, facts proved that Huawei was clean.

On July 28, 2004,[2] Huawei, Cisco, and 3COM submitted an application to the Marshall District Court of the State of Texas requesting an end to the suit. The court thereupon issued an order putting an end to Cisco's suit against Huawei. This had been the largest dispute between China and the United States over intellectual property to date and it was ultimately concluded by a handshake agreement.

Looked at objectively, this multinational suit initiated by Cisco did not in fact achieve its strategic objective. It did not result in putting an absolute end to Huawei's attack on North America. On the contrary, not only did Huawei prove that it was blameless, as determined by international standards, but the case served to amplify its international reputation. Huawei also paid a heavy price, however.

By "private agreement," Cisco was given certain protections on American turf. Influenced by this and other factors, Huawei's advance was now made more difficult. The company had no alternative but to moderate its attack on North America. Moreover, it now redirected the focus of its major battles in the direction of Europe and other overseas markets.

---

[2] The author notes that this was the date in the United States (it would be the following day in China).

CHAPTER THIRTEEN

TRADING LAND FOR PEACE

*Huawei is still small and still weak. We are not yet ready to go
directly up against international "friendly companies," so we
must hide our true capacities and bide our time. We must learn
from those we invite in as guests, trade off some turf for peaceful
relations. We would rather let go of some markets and some
interests, cooperate with friendly companies, partner with them,
than go against them. We then create new space in which we can
coexist and both enjoy the benefits of the value chain.*

—REN ZHENGFEI

Huawei's landing on the soil of North America had been powerfully
rebuffed. Nevertheless, Huawei garnered an enormous harvest from the
experience. The truth of the matter was that the mantra of "high quality,
excellent service, low cost" was unbeatable in business competition. Still,
the "private agreement" was protected and upheld on American turf. This
international principle could not be altered. Landing on U.S. turf now
became extremely difficult for Huawei, while it was even harder and more
brutal in Europe. In Europe, a crowd of telecom giants operated at will,
including Ericsson, Siemens, Nokia, and Alcatel.

Ren Zhengfei was highly aware that Huawei could not rely solely
on its own strength if it intended to cross the turbulent Atlantic successfully and lodge its forces on European soil. It was going to have to take
advantage of a "major battleship" or perhaps even a "mother spaceship"
if it was to land successfully on the opposite shore. That is, if Huawei

was going to emerge on the international stage, not only would it have to "hide its true nature and bide its time," and "pull in guests in order to learn from them," not only was it going to have to "trade some turf for peaceful relations," but it was going to have to form an international alliance of forces. It was also going to have to internationalize its sales channels and its brand.

Difficulties in Forming the Alliance

Ren Zhengfei's strategy of "borrowing someone else's ship to go abroad" was a key maneuver. It was perfect for a Chinese company with weak brand influence that sought to open up international markets.

In the early 1980s, when Motorola was entering Japanese markets, it was confronted with tremendous trade barriers and found it hard to move an inch. Motorola only cracked open the Japanese market after operating a joint venture with Toshiba for a number of years. Both parties developed, produced, and marketed products together. At around the same time, multinationals were entering China's markets by using the same joint-venture methods in order to accomplish their offense strategy of trading technology for markets.

In the initial period of opening up international markets, Ren Zhengfei led the way in using joint ventures to open up markets in Europe and Latin America. He tried such joint ventures first in Russia, Yugoslavia, and Brazil. Given the minimal influence of Huawei's brand in international markets, and also given the cruel realities of one defeat after another, Ren Zhengfei made the decision to accelerate this tactic of setting up alliances.

In 1998, Huawei held discussions with three major multinationals about cooperative endeavors, namely Lucent, Nortel, and Ma-ke-ni. With all good intentions, the idea was to borrow their brand influence and their channels to open up European and American markets, so that both sides could enjoy the benefits provided by the value chain. These three companies did not regard Huawei's earnings as in the same class as theirs, however. Since their annual income was on the order of several billions of dollars, they refused. Meanwhile, companies that voluntarily proposed cooperation to Huawei had their own ulterior motives. For example,

Motorola did not own program control switching technology, so wanted to make up for its own deficiencies through working with Huawei. It then would be able to move into China's own GSM market. Cisco also held business negotiations with Huawei about cooperating. Its condition was that Huawei would give up R&D on high-end routers. Cisco would pass over the production business of low-end routers to Huawei. Ren Zhengfei naturally was opposed to a form of cooperation whereby Huawei was merely "doing the hard labor" on behalf of Cisco.

Forming alliances was clearly going to be difficult, but Ren Zhengfei, strong willed and ambitious as ever, did not change his mind. He continued to reiterate the strategy in internal meetings. He said that developing cooperative partnerships was not done in order to gain short-term benefits but rather to smooth the road for Huawei's long-term growth objectives. Moreover, Huawei absolutely must maintain control shares in such alliances. Although Huawei's expansion into overseas markets was extremely difficult over the next several years, Ren Zhengfei did not in any way abandon this principle. By 2001, strategic alliances in which Huawei played a key role now began to turn for the better.

Joining Forces Both Vertically and Horizontally

The company 3Com, a contraction of "Computer Communication Compatibility," had invented certain kinds of routers and Ethernet technology at an early stage. Since it remained steeped in its former glory, however, it missed the boat as the Internet developed. As Cisco emerged, 3Com abruptly woke up from its dreaming only to discover that not only had Cisco occupied its high-end markets but smaller Silicon Valley companies were now nibbling away at its low-end markets. It only took half a year for the company to go from being the industry leader to being a company that was quickly moving in the direction of major losses. All of this was painful enough for 3Com to realize that it had to find a partner if it was going to regain its ruling throne in the global network business. This partner had to be powerful in both capital and technology if it was going to go up against common enemies in the international arena. After evaluating all options, 3Com chose to work with Huawei.

In mid-November, 2001, a delegation led by 3Com's chief technical officer came to visit Huawei. With great sincerity and candor, the team introduced the prospect of the two sides setting up a joint venture. The person receiving this visit at Huawei was Guo Ping, vice president at the time. After listening to this proposal, he remained quiet and simply treated the visitors to a banquet. He noted that "others have to look more deeply into the proposal after our lunch." Since the 3Com delegation did not yet know Huawei's real intentions, they were disappointed and had less appetite than they might otherwise have had for the banquet. Nevertheless, they became quite pleased when they discovered after the banquet that "other people" actually meant Ren Zhengfei himself.

What then transpired was quite simple. Negotiations on the details of the joint venture began secretly between Ren Zhengfei and 3Com's CEO, Bruce Claflin. The infringement suit that Cisco brought against Huawei lay down a perfect red carpet for this multinational cooperation. It also initiated a number of long-lasting cross-national marriages for Huawei.

On March 20, 2003, Huawei and 3Com made the announcement in the United States of the formation of a joint venture called H3C.[1] On August 29 of the same year, Huawei and Siemens announced the formation of a joint venture called Dingqiao Telecom. The announcement was made in Beijing. On September 22, Huawei, NEC, and Matsushita announced a three-way joint venture in Shanghai called Yumeng Telecom.

In the short space of half a year, Huawei had set up three joint ventures with four world-class companies. This was extremely rare in the history of business around the world. Moreover, the headquarters of all of these were located within China.[2] Not only did this form of joint venture reflect Huawei's dominant position but it also created extremely advantageous conditions for Huawei to open up international markets.

Although 3Com did not own sufficient technology to challenge Cisco, as a founder of modern networking equipment it had tremendous

---

[1] This is also known as Huawei-3Com Co., a Hong Kong-based joint venture. In 2007, 3Com acquired 100 percent ownership of H3C. Huawei later (2007) agreed to provide minority equity financing for a purchase of 3Com by Bain Capital. This fell through in early 2008 due to U.S. concerns about cyber security.

[2] Hong Kong is therefore considered to be within China for purposes of this statement.

influence in America. At one point, its market value was on a par with that of Cisco. Moreover, 3Com had a network of 50,000 agency channels that were globally distributed. After H3C was set up, not only could 3Com shift its R&D Center to China, thereby reducing its R&D costs and also production costs, but it could take advantage of Huawei's technology resources to restore its former glory. This was undoubtedly attractive and indeed was the reason 3Com put USD 165 million into the venture plus its 50,000-strong channels. It agreed to hold 49 percent of shares in the joint venture.

As for Huawei, it merely invested its product lines in digital telecom, and thereby gained 51 percent of shares. Not only did Huawei derive the invaluable experience of learning how world-class companies handle things in real battles, but Huawei found its name recognition catapulted around the world. Even more importantly, by taking advantage of 3Com's brand influence and its channels, Huawei could now directly begin to force its way into the long-dreamed-of North American markets.

Huawei's joint venture with Siemens was even more strategic. Back in early 1994, Ren Zhengfei just emerging from a proverbial reed hut,[3] had already started the slogan about "dividing the world into thirds." The primary contender for one of these thirds had been Siemens. This initially made it somewhat awkward when the two opponents sat down to discuss investing USD 100 million in a joint venture. They were now talking about joining forces in developing and producing TD-SCDMA. Nevertheless, under the persuasive power of mutual benefit, both sides quickly broke the ice and moved toward agreement.

Siemens had 11 percent of the patents for TD-SCDMA. However, since the international standards for this technology had been formulated by the Datang Company, Siemens hoped very much to cooperate with Huawei in order to enable it to share in the market, have its own bowl of porridge so to speak, when it came to eating the TD-SCDMA market.

As for Huawei, the company's advantage in the 3G age had been expressed in terms of WCDMA. Now, with TD-SCDMA, its advantage was reduced. Cooperation with Siemens was perfect in helping it make

---

[3]This is the equivalent of "a garage in Silicon Valley."

up for its own deficiencies. Moreover, Siemens was the core agent for Cisco in distributing digital telecom products. After Huawei and Siemens linked up, the emotions of the two were much more in sync. Huawei provided Siemens with extremely beneficial business terms in asking it to represent Huawei in distributing products in Europe. As soon as the flinty-eyed Siemens saw the profit margin in these figures, Huawei's sales of digital telecom products in Europe quickly broke through USD 100 million.

In Japan, the tri-partite joint venture of Huawei with industry giants NEC and Matsushita had total registered capital of just USD 8 million. Huawei, meanwhile, held only 6 percent of shares. Nevertheless, NEC's chairman, Nishigaki Koji, personally served as chairman of the board of this company, while Matsushita's chairman served as CEO. Ren Zhengfei served as a board member, which clearly demonstrated the importance of the joint venture to all three.

For NEC and Matsushita, the 3G end-user business was a core business. Through close cooperation with Huawei, these companies not only could use the low-cost advantage that Huawei provided in R&D to strengthen their competitiveness against Nokia and Sony Ericsson, but they could bundle their products with Huawei's 3G equipment. As much as possible, they could achieve their strategic objective of expanding into both China and international markets.

As for Huawei, through cooperation with NEC and Matsushita, the company not only could directly upgrade the technological content in its own end-user cell phones, but at the same time it could break into the Japanese market via NEC's and Matsushita's channel influence, so as to enter the optical network, digital transmission, and 3G product markets there. Facts were later to prove that the reason Huawei was able to break through the highly exclusive 3G market in Japan in advance of European and American telecom giants was that it put major effort into developing the end-user product joint venture company Yumeng at a critical time. This was in addition to Huawei's own decisive advantage in technology. The combination played a key role in helping Huawei "push the wave."

If You Intend to Conquer the World,
You Have to Make Friends with Enemies

There are no eternal enemies and there are no eternal friends. There are only eternal interests. This wisdom, as summarized by the 19th century British Prime Minister, the 3rd Viscount Palmerston, was confirmed yet again as Huawei proceeded with its international strategy.

Huawei had no advantage in technology in its early period since it was a "tech follower." To gain a foothold in markets, all it could do was compete on the basis of price. It then sought to expand market share—this was simply the most realistic and feasible plan. In the course of competing in overseas markets, Huawei found that the telecom giants in Europe and America could not use price advantage in their favor but they could and did use other highly effective weapons. Specifically, they set out land mines to block competitors, in the form of patents. Huawei's encounter with Cisco belonged to this class of maneuvers.

If Huawei was to become a world-class company, without compromising itself in any way, and if it was to achieve long-term growth in international markets, it had to break through this dividing line between friends and enemies. To do that, it had to give up some immediate interests and trade those for long-term strategic interests. To create a better environment for itself, it had to cooperate.

To this end, in February of 2003, Ren Zhengfei made the following clear declaration after he had listened to the year-end sales meeting, with reports from overseas frontline marketing staff. "Huawei is still small and still weak. We are not yet ready to go directly up against international friendly companies, so we must hide our true capacities and bide our time. We must learn from those we have as guests, trade off some turf for peaceful relations. We would rather let go of some markets and some interests, cooperate with friendly companies, partner with them, than go against them. We then create new space in which we can coexist and both enjoy the benefits of the value chain."

He went on to say, "As we open up overseas markets, we should not engage in price wars. We must find win-win solutions with friendly partners. We cannot mess up markets and we must avoid having western companies gang up on us. Instead, based on our own hard work, high-quality products, and excellent service, we must achieve customer support and acknowledgment. We cannot damage the profitability of the entire industry just for our own small bit of sales. We absolutely must not be the one who destroys market order."

With respect to how to deal with the "patent land mines" issue, in early 2004, as he was visiting the European headquarters of Huawei, Ren Zhengfei told employees, "Huawei is still a foot soldier when it comes to comparing us with the international leaders in the industry. They have been amassing patents for over one hundred years. There is simply a major gap between us. Because of that, we must actively learn from these stronger forces. We must respect their position as market leaders. We must engage in competition but on an orderly basis, so that we can achieve win-win situations."

In order to express the sincerity of this approach, Ren Zhengfei constantly reiterated this theme inside the company. Huawei had to construct a cooperative environment in which to operate. It had to strive for more cooperative opportunities with competitors. Ren Zhengfei then emphasized that employees were not allowed to even use the term "competitors." Instead, such companies as Nokia, Siemens, Ericsson, Lucent, Alcatel, and other telecom giants in Europe and the United States were to be referred to as "friendly companies." In tenders for projects in markets in which Huawei held an advantage, such as Russia, Asia, and Africa, Huawei was to use a strategy of "jointly enjoying the value chain with friendly companies." It was to join together with such friendly companies as Alcatel, Lucent, and Northern Telecom in partnering on bids. In other markets, it was to merge its interests with others by bundling sales and other ways to "join hands to get orders." The idea was to achieve ever more fused relationships.

Going from Folk Dancing to Dancing the Tango

Although Huawei lost some profit margin by these methods, it also enhanced its long-term survivability.

In around 2003, Huawei had been excluded from some major high-end projects in European and American markets. The moment Huawei made a bid, other telecom giants would join together to raise the technological bar. Through such measures as setting up "patent-strewn mine fields," they would keep Huawei out, then divide the value chain among themselves. Now, as cooperative relations with Huawei constantly improved, not only did these telecom giants join hands with Huawei in making bids in order to provide both OEM products and services, but the mine fields that they had so painstakingly constructed were gradually dismantled.

Starting in the second half of 2003, Huawei signed licensing agreements with Ericsson, Nokia, Siemens, Nortel, and Alcatel, which allowed it to use patented technology on a fee-paying basis. This allowed Huawei to achieve the credentials for being an international competitor prior to when that might otherwise have happened. It also opened out "living space" for the company. In 2004, Huawei's fees for using other people's patents came to USD 30 million, but in that same year the company realized USD 5.6 billion in sales. In 2008, Huawei spent more than USD 200 million on patent-use fees, but its sales came to more than USD 20 billion. Huawei would not have been able to achieve these figures if it had not concluded patent agreements. It certainly would not have been able to grow as swiftly as it did.

In order to become a world-class company in the true sense, Huawei made broadly based friendships through the strategy of "exchanging turf for peaceful relations." It constructed and then expanded an amicable "space" into which it could grow. Internally, it also continued to put greater effort into its own inventions and patent filings.

In addition to granting high bonus awards to inventors, it established a rule whereby patents were to be filed in the name of the inventor. This proved extremely effective as a way to motivate creativity.

In 2000, Huawei only submitted one international patent application to the World Intellectual Property Organization. By 2005, the company submitted 249 applications and leapt to 37th in the ranking of global patent submissions. For the first time, it surpassed its old opponent, Cisco. By 2008, Huawei had become the number one patent filer in the world, with 1,737 international patents. Since it now possessed a sufficient number of core patents to give it equal standing with other multinationals, Huawei was asked to serve in leading positions in the more than 180 international telecom organizations and standards organizations in the world. It became chairman or vice chairman, director or chief editor. It organized, chaired, or participated in the planning for, design of, and formulation of international standards. Moreover, it carried out interlocking patent exchanges with various "friendly companies," which created ideal conditions for its entry into European markets.

One example is DSLAM, which was the unique invention of Alcatel. Distributed base stations, however, were Huawei's invention. In order to achieve the fastest possible entry into these two spheres, both sides signed an agreement in 2004 for exchanging patents. As a result, Huawei's sales in the DSLAM arena quickly leapt to second place in the global ranking. Between 2008 and now, moreover, Huawei has surpassed Alcatel in this area and been ranked number one in the world.

As Huawei's "right to a voice" in the arena of international patents constantly improved, Ren Zhengfei's friendship with the CEOs of Alcatel, Siemens, and Nortel has also grown. Forms of cooperation have gone from such things as joint branding of products and exchange of patents to more strategic levels, including joint development of core technology and joint strategic investments that are oriented toward the more distant future. This even extends to closer cooperation between Cisco and Huawei. When Cisco's chairman, Chambers, visited Ren Zhengfei in Shenzhen in December, 2005, in addition to expressing the sincere desire to exchange the weapons of war for an olive branch, he expressed the opinion that the two sides should carry on more in-depth forms of cooperation.

Through this strategy of "exchanging turf for peaceful relations," Huawei was able to set up multi-tiered cooperative systems on a global scale. From simply purchasing the right to use other people's patents, the company now sat on an equal basis with telecom giants. In this process, Huawei, which Ren Zhengfei had always described as "a small workshop," was now standing on the international stage working with world-class companies as partners. It was also being forced to shuck off its straw sandals and put on proper leather shoes. It was taking the very pleasing step of going from folk dancing to learning the tango.

Moreover, as time went on, it was more and more adept at dancing this tango.

Hit with the Most Brutal "European-style" Experience

British Telecom, known in the industry as BT, had always been known as a venerable operator that focused on technology and that had an innovative spirit. Its business operations covered more than 170 countries and regions. It was at the absolute apex of world-class companies.

When Huawei's people first had contact with British Telecom in the early days, BT knew that Huawei had developed program-control switches but it did not believe that Huawei could come up with any real technological innovations. Once Huawei used its "soft exchange equipment" to break into Evoxus company, whose control shares were owned by British Telecom, the British still expressed doubts about Huawei's products. After hitting up against several walls, Huawei's people finally began to realize that they would have to go through a stringent certification process if Huawei was to win BT's trust and become a supplier. BT's certification process was known for being extremely strict and meticulous. Not only did this take time, but it had to cover a broad range of equipment. It was regarded as the most authoritative certification in all of Europe. This even allowed many manufacturers to be exempted from further inspection from operators in other countries, once the products had been certified by BT.

In early 2003, BT issued its "Plan for a 21st century network," and published this plan worldwide. Its aim was to ensure that any user could connect via high-speed broadband with any end-point and in any place, via any end-user device. The user could smoothly enjoy comprehensive access to voice, images, data, and wireless telecom, all through one device.

This was a massive project. It would take five years to build and cost more than 10 billion British pounds. Not only was this the largest tender for a telecom project ever issued in the world, but its far-sighted vision and audacity amazed the world. If a supplier could become part of this British Telecom project, not only would that mean that it had enormous capability as a world-class provider of products and solutions to customers, but also that it would be opening the gates for itself into an exclusive global club.

The opportunity that Huawei had long hoped for and anticipated had finally arrived.

In order to join the British Telecom team successfully, Huawei specifically set up a "certification leading small group"[4] that was led by the Chairman of the Board, Sun Yafang. It included the Deputy head of R&D, Cai Feimin, the head of Human Resources, Xu Lixin, and Senior Vice President Lu Ke. This team began detailed preparations as per BT's certification requirements, which lasted for three months. The requirements included eighteen major sections and more than 200 subsections that covered R&D, markets, supply chain, production, finance, and also human rights. However, in the face of people who were supremely experienced in dealing with international certification, Huawei's team was as unprepared as school children who had not done their homework. Inspections soon revealed the company's shortcomings. In a report that they titled "Regarding Huawei's recent personal experience," Fei Min and Xu Lixin summed up the experience. "When BT experts met with our most experienced work-flow experts, quality experts, and senior management, they first asked, 'Who among you all sitting here can tell me which five most important issues affect Huawei's ability to deliver high quality products and services to the customer, in terms of the end-to-end sequence of processes? Which five are most in need of solutions?' Not one of us could answer that question!"

Another problem was quality control, called "supervisory management over workflow procedures." As one of Huawei's team later explained, it was quite awkward for the company when a clear violation of safety procedures occurred under the very eyes of BT's certification experts. After five excruciating days of being tested and evaluated, British Telecom experts finally provided Huawei with the results of its evaluation. The company's manufacturing plant, production lines, and other such basic

---

[4]Chinese government policy is often made by what are called "leading small groups." Huawei has therefore echoed that procedure in its internal structures.

infrastructure were world class. However, in terms of R&D, the supply chain, and capacity to deliver end-to-end business, and critical standards relating to these, there was still a certain disparity between Huawei and the standards of world-class companies.

Huawei people were powerfully affected by this conclusion from British Telecom. The three senior people, Fei Min, Xu Lixin, and Lu Ke, evaluated the situation as follows. "Some of our colleagues do not accept this. They are not convinced. They say, 'Haven't we already adopted the world's most highly regarded integrated R&D management procedures? Haven't we taken the international lead in implementing CMM systems? Weren't we successful in obtaining certification from a number of other countries?' What we say to this is that if you put Huawei up against other international competition, in that context we are immature and naïve. You have to regard Huawei as a company that is going from local to international, and we are not there yet."

"This 'personal experience' gave us an opportunity to firm up our understanding of what an international mainstream operator expects and wants from a potential partner. It made us realize what a core supplier has to be. The process as well as the results allowed us to see our own shortcomings in terms of management, processes, product development, service provision, and a host of other things. These disparities now mean that we must go forward with very well-placed steps as we continue to improve ourselves."

One of the most valuable parts of this hard-earned experience was that Huawei people now recognized that "overseas markets reject opportunists." They now had a profound appreciation for "becoming stronger by being humbled." In the following two years, as BT continued to test Huawei's innovative capacities and its ability to do customized production, Huawei's responses were faster than any other electronic equipment firm. The company was able to deliver appropriate solutions proposals within a shorter time.

Over the course of these two years, Huawei invested several billion RMB to implement systems that complied with BT's certification. As per BT standards, it set up more than 200 regulated processes, complying with international standards, to do with quality control, delivery, environmental protection, human resources, and customer satisfaction.

As a result, the company then did pass full certification. Huawei is now no different from Cisco, Ericsson, Alcatel, or other multinationals when it comes to production capability, standards of delivery, R&D deployment, and procedures. During this time, moreover, Huawei transformed itself and fulfilled Ren Zhengfei's dream of taking the company from being a "small workshop" to a world-class multinational.

In 2005, one of the world's most meticulous and formidable authorities on telecom certification made the following declaration: "Huawei's success has not come by accident. Instead, as an international firm, its ongoing growth is safeguarded by modern management systems and by the very best of global telecom technologies." The Englishman came to this conclusion after completing his comprehensive certification of the company. He then headed home, after spending the Christmas of 2005 in China.

Equal Status in the World 500 Club

On April 28, 2005, British Telecom announced the names of the eight companies that were selected to become suppliers in the "Plan for a 21st century network," the winners of the tender for more than 10 billion British pounds. Huawei was included in the list, together with Ericsson, Lucent, Cisco, and other multinationals. What most amazed people was that Huawei was the only company to be chosen as supplier for two different major spheres of activity, network accessing and optical transmission.

After the list was published, mainstream media in Europe and America reported on the painstaking process of BT's certification of Huawei. England's *Financial Times* noted that "China's Huawei is in the process of rewriting the rules of the game for the globe's telecom industry."

Huawei had not just been given a ticket to ultra-large orders and entry into an exclusive club. Something of greater significance lay behind these reports and accolades. At the time, the world looked on Chinese products as inferior—Huawei's rise destroyed this preconception. It overturned the dismissive attitude with which western powers had viewed China ever since the Opium Wars.

At the end of November, 2005, more astonishing news came from England. Huawei had signed a framework agreement with the largest mobile operator in the world, Vodafone, to become its strategic supplier. It was to provide complete systems products, everything from cell phone terminals to broadband access to optical networks, including 3G. Prior to this, the mysteriously ranked list of strategic suppliers to Vodafone had included four different companies for different products, Ericsson, Nokia, Siemens, and Lucent. Now, Huawei was actually assuming responsibility for all of them.

All of this news, from British Telecom to Vodafone, meant that Huawei now had the ability to help world-class customers upgrade their own all-round competitiveness. Because of this, doors began to open to Huawei around the world, in Germany, Italy, Spain, Poland, Portugal, Japan, and South Korea.

In March, 2003, Neuf announced that it had selected Huawei to build a broadband access network throughout France and that it would be doing an on-site tour of Huawei's headquarters in Shenzhen. Some journalists were invited to go along. They were astonished by what they saw. One journalist even cut his trip short and hurried home to spread the word. His headline: "The rise of Huawei will spell disaster for multinationals." He warned large telecom companies that they were in for a serious challenge from this Chinese company. This warning was not given much credence at the time, however. It was seen as insignificant among much more important items of news. Five years later, once the reality of the nightmare had fully awakened Europe's telecom giants to the dangers, they discovered that Huawei was emerging in their midst, already fully formed.

CHAPTER FOURTEEN

HUAWEI'S EMERGENCE ON THE SCENE AND RESULTING CONFRONTATIONS

*When a market economy is overextended, the situation is
like a "wringing something to death." What does that mean?
It is like wringing out a washcloth. There is still room for
competition when you can still wring out water. When you can't,
that means the other enterprise is finished. The best situation is
when the washcloth is wrung dry, but the washcloth itself is still
in one piece.*

—REN ZHENGFEI

Marconi is England's largest manufacturer of telecom equipment. It was founded in London in 1897 by the "father of wireless," Guillermo Marconi. His invention of the telegraph, the transmitter-receiver, or broadcasting stations, and radio, not only began the great current of today's wireless telecommunications, but also profoundly affected the advancement of humankind.

In the late 1990s, when Marconi announced that it was moving its forces into the telecommunications industry, its share price soared to a height of 350 pounds. What reaches an extreme must always return, however. At the time of its greatest glory, Marconi made a fatal mistake. In 1999, when the IT industry was flourishing, the company spent a large amount of money to buy several optical transmission manufacturers. Its core business became misaligned with what should have been the correct orientation. In the IT winter that followed on the heels of this decision, the company sustained repeated setbacks before becoming mired in debt.

Heavily wounded, Marconi nevertheless worked hard at strengthening its after-sales servicing and expanding its product line. It lowered its management costs in order to step onto the path to recovery. Fate was soon to come down on it again like a landslide, however.

On April 28, 2005, as the eight victors of the 21st-century project were celebrating their success, Marconi was in dire straits.[1] Two days after the news about the 21st-century project was announced, Marconi's share price fell by 57 percent. Its market valuation evaporated 280 million pounds in that one day. In the midst of the howls of the British public, European media now announced another bit of explosive news: "Marconi, on the brink of bankruptcy, intends to sell its core assets. The most likely buyer is the dark horse who just beat out the competition in British Telecom's tender, namely Huawei."

What Should Huawei Do in the Midst of a Massive Wave of Mergers?

Historically, Marconi's products and Huawei's products emphasized different things. The two were not competitors in a strict sense, so in 2001 they started joint branding on some products. Their association was fairly amicable. In January 2005, prior to the announcement by British Telecom of its list of winners, Ren Zhengfei flew to England and met with Mike Parton, chairman of the board of Marconi. The two formally signed an agreement for cooperation in which each would agent the other's products and they would jointly develop international markets. Once Marconi failed to be chosen in the British Telecom bidding, however, and its stock began to spiral downward, blame and recriminations were ubiquitous. The board of directors of the company was particularly dejected as the shares continued their steep descent. After weighing all the options, they came to a conclusion: it would be better to be bought out altogether than to become even more mired in debt.

---

[1] Literally, "Marconi had fallen behind even Sun Shan, the one who came last in the imperial exam."

The board therefore expressed this idea in an exploratory way to several companies, including Huawei. Ren immediately dispatched a vice president to go talk with them, someone fully versed in multinational acquisitions and capital operations. After several rounds of negotiations, Huawei formally made an offer of 682 million Euros for Marconi.

Once this news was officially confirmed by the board of directors of Marconi, Ericsson entered the fray with a higher bid of 1.2 billion Euros, and Marconi accepted this bid.

In December 2005, in one leap Ericsson therefore became the largest telecom equipment manufacturer in the world. Good things do not last forever, however. On April 2, 2006, Alcatel and Lucent announced in Paris that the two were formally merging. On June 19, 2006, Nokia and Siemens then also announced that the telecom equipment departments under the banner of each company would be merging to form a new company.

When this news got out, the entire industry was in an uproar. According to the latest ranking of telecom equipment manufacturers, ranked by operating revenues and put out by the International Telecom Union, Alcatel and Lucent now ranked first in the world with postmerger revenues of 19 billion Euros. Ericsson ranked second, with revenues of 16.2 billion Euros. Nokia and Siemens ranked third, with revenues of 15.8 billion Euros. Huawei merely ranked eighth, with revenues of 3.6 billion Euros. It came after Nortel, Motorola, Cisco (the telecom business part of the company), and NEC.

In the short space of half a year, the picture had changed. Companies that had formerly been dire enemies for years now became part of family. Such a swift wave of mergers was extremely rare in the history of business. After these giants merged, they had more customer resources and technical resources to share with one another, but they were able to provide much lower-cost products and services to customers due to such things as firing redundant personnel and redundant departments and cutting back on excessive management expenses. What's more, after merging, these companies had more time and energy to focus on what they called the "China threat," by which they meant such companies as Huawei and Zhongxing. They were able to adopt more forceful measures to keep these companies from further expansion.

Clearly, this shuffling of the cards indicated that Huawei's price advantage would slowly be eroded. It also meant that the gap between Huawei and other telecom giants was now suddenly greater. Moreover, in the process of acquiring Marconi, Huawei had come up against strong intervention by both British and American governments. Huawei's "march toward globalization" was now more difficult and more complex.

Positioning Determines Status

Traditional telecom markets had been heavily impacted by the vigorous growth of Internet technologies. Meanwhile, as Huawei's price wars wreaked havoc, margins were pared down. It was only rational for giants in the industry to merge in order to strengthen their product lines and improve their customer-line advantage. This was simply the appropriate solution. So far as Huawei was concerned, the situation also presented advantages. Despite its isolation and lesser strength, Huawei could develop at a fast pace even in the face of the wave of mergers if it could manage to grab hold of more complete product lines and maintain excellent customer lines.

Positioning determines status. Based on this line of thinking, one of the most important product-line measures that Huawei adopted, in late 2006, was what it called a "Five-year development plan for total IP." That is, via IP platforms, Huawei would integrate 2G, 3G, and 4G technologies into one network and overturn the traditional mode by which one had to buy three sets of equipment for three different networks. Not only did this save customers a large amount of money, but also it solidified Huawei's competitive position. It gave it the ability to direct the action in the age of both 3G and 4G.

In terms of customer-line measures, one of the most important things Huawei did was to set up a second-tier sales and marketing department on a global basis. On top of its existing structure, which was both strategic and market oriented, it now added this to its 17 R&D centers around the world, as well as its 26 innovation centers and 22 regional departments. To ensure adequate human resources for this effort, Huawei began implementing a massive five-year recruitment plan. By the end of 2006, Huawei employed more than 60,000 people. By the end of 2008, it employed 87,500, and by the end of 2012, it employed 130,000. A small percentage of these new hires remained at headquarters to handle administrative functions, but the great majority were sent out and around the world to various sales and marketing departments. They had dual functions. In addition to collecting, understanding, and evaluating customer needs, and using that information to propel R&D and the development of emerging products, they promoted Huawei's "three-in-one" product-solutions plan to customers. That is, they promoted three networks combined into one.

As the combined sales and R&D network was extended around the world, tying in frontline customers, Huawei put out a new slogan: "Whoever is closest to frontline markets gets to become the leading class." In interacting with customers, the sales and marketing departments followed the principle of "universal customer relations" that had been practiced in the domestic market. That is, they established excellent relations at all levels of the customer. All parts of the company were therefore highly motivated to make sure Huawei got a "vote of approval" when competing for tenders. Huawei's customer-line advantage became extremely well entrenched.

A Customer-Centric Approach

The drafting of *Huawei Basic Law* had begun in 1996. In the decade between 1996 and 2006, Huawei went from being a small workshop to being a global telecom equipment supplier. The role that *Huawei Basic Law* played in this process was invaluable as a way to "escort the Emperor" and guide the ship safely. As patterns of competition changed, however, and as Huawei's international status was transformed, the original mission of the company as stated in *Huawei Basic Law* could no longer satisfy the company's growth needs. The mission, objectives, and core value system of the company now needed to be modified.

In the second half of 2005, Ren again called together the six eminent professors who had formulated the original *Law*. They combed through and reevaluated the core values and positioning of the company and, after six months of internal discussions and revisions, they confirmed a new core value statement. Ultimately, this came down to regarding the customer as central, that is, changing to a customer-centric approach. The objective of the company now became "to serve the customer as the sole reason for Huawei's existence." This allowed for an organic integration of Huawei's vision, its mission, and its growth strategy.

To externalize the new core value statement, and make it visible, Huawei now created a new image for itself as a technology leader. On May 8, 2006, the company revised the corporate logo it had used for the past 18 years. Instead of "15 rays of sunshine," it now had a beautiful chrysanthemum with eight petals. In international markets, this symbolized the fresh and outstanding products and technology that Huawei provided to customers. At the same time, it indicated a willingness to work together with partners and "friendly companies," to construct a congenial business environment in which all could grow together.

Ren also changed the corporate logo to drive home a deeper meaning. He wanted China's chrysanthemum to bloom in every corner of the world on an enduring basis. This required not only compatibility, cooperation, and openness, but also a longstanding willingness to work extremely hard toward the same goal. Ren wanted all employees to keep firmly in mind that the customer was central to everything Huawei did, not only in the immediate present but into the future. The "customer as central" had to be the ultimate principle and value judgment by which all carried out their work.

Molding a Corporate Culture in Which Everyone Rows Together

As Huawei revised its vision, mission, strategy, and logo, unmistakably putting out a whole new image, Huawei's long-standing corporate culture inevitably also had to move forward with the times.

In the early years, Huawei espoused a "wolflike culture" that used high-pressure means to achieve results, including compensation mechanisms that rewarded achievers. To the company's founders, who were born in the 1960s and 1970s, this culture played a powerful role in motivating people. As Huawei became more international, however, and as the competitive environment changed, this corporate culture now faced severe challenges.

By the first half of 2006, Huawei had become a multinational with more than 60,000 employees, among whom 80 percent were born after 1980. Their aspirations were more individualized and more human-oriented. They were more intent on realizing their own personal values. Moreover, large numbers of non-Chinese nationals had now been brought into Huawei, while tens of thousands of Chinese nationals had been sent abroad. Huawei was therefore facing the many challenges posed by cross-cultural management. If it was to become an international company, it not only had to maintain its wolflike culture in the sense of working together as an indomitable pack, but also had to realize a high degree of synchronized professional spirit. Like a crew team in the Olympics, it had to row in unison. Ren considered how to accomplish this, what cultural "vehicle" to use to push forward Huawei's strategic transformation. In 2005, he found his answer as he watched a dance program on television.

The program was over in a brief 5 minutes and 54 seconds, and it had no music accompaniment. Yet 30 young women performed a dance in precise unison with one another, reaching a realm of artistry that athletes would have found difficult to achieve. Broadcast on national television, this touched the hearts of millions of Chinese. The women remained tranquil and unaffected by the applause that followed, however, for they were completely unable to hear.

Their performance and their attitude were precisely what Huawei's new culture was striving to achieve. It sought excellence, was highly disciplined, worked together as a team, yet was unaffected by adulation and glory. These were value standards and a professional spirit that met the demands of Huawei's new culture. They embodied the idea of "everyone's oars rowing in unison." As Ren was later to say in a speech at a training session for regional finance managers, "Someone asked me, 'How would you describe a visual image of Huawei's culture?' I thought about this for a long time. After I saw those girls dance, despite their being deaf and mute, I recognized the extraordinary professionalism of their performance. In a sense, all their oars were rowing in unison, and I realized that this must be the very spirit behind Huawei culture."

This began to symbolize Huawei's new culture: all oars rowing in unison. The new standards by which performance was evaluated, the new professional characteristics, were now encapsulated in various phrases that promoted team cooperation and compliance with natural laws of harmony. Between 2006 and 2012, the number of Huawei employees went from 60,000 to 130,000. Sales revenues went from USD 10 billion to USD 30 billion. During this period, performance evaluations still adhered to the "five grades and double-channel" system as the decisive consideration, but "rowing in unison" played an enormous role in bridging cultures and getting the company to perform as one entity. It pulled people with different nationalities, beliefs, and value outlooks into one team, working toward one goal. As a result, Huawei's overseas markets surged ahead in a second spurt of growth.

In 2006, Huawei's operating income of USD 8.5 billion surpassed that of Motorola, NEC, and the telecom part of Cisco. The company entered the ranks of the top five telecom equipment companies in the world. In 2007, Huawei moved into fourth place with operating income of USD 12.6 billion as it surpassed Nortel.

Rescuing Profitability

The accomplishments were obvious. Nevertheless, Huawei was paying a price for its success. According to the annual report published in April 2008, Huawei's operating profitability fell from 19 percent in 2003 to 7 percent in 2007. In the same period, its net profit margins went from 14 percent to 5 percent. Prior to 2005, the company had consistently maintained a debt-to-assets ratio of less than 50 percent. By 2007, this rose to 67 percent. Meanwhile, in the course of high-speed growth, Huawei's accounts receivable rose dramatically to varying degrees in different places. Cash flow now presented the potential for very considerable operating risk.

In the second half of 2007, after taking inspection trips to dozens of Huawei branches in Europe, Africa, and Latin America, and combining what he learned with the operations and market competition in these places, Ren proposed a transformative way of thinking about the company's business. He called this "seeking to create order in the midst of turmoil, and seeking to release turmoil in the midst of controls." For five years, a previous slogan in the company had been kept rather muted, namely, "cutting the feet to fit the shoes." This now resounded throughout Huawei once again.

In 1998, Huawei had worked with IBM to figure out what business procedures to bring into the company as a part of revamping its financial and accounting systems. Since management of financial matters and also customer affairs involved business secrets, both sides ultimately decided to launch Integrated Product Development (IPD) and Integrated Supply Chain (ISC) work-flow procedures, first, as an initial reform. Per the recommendations of the IBM consultants, Huawei worked with well-known international consulting firms to implement unified financial systems that incorporated four different aspects: unified financial categories, unified sequences of work-flow processes, unified codes, and unified monitoring. This represented a transitional step in the overall restructuring of financial management. In September 2003, Huawei also began to implement three revamped systems to put it in alignment with international practices. Those related to accounting, budgeting, and auditing. These reforms played a critical role in controlling Huawei's costs and expenditures and in improving both profitability and accuracy. However, when it came to controlling operating risk and putting the company in line with international financial and accounting systems, they still left Huawei at a far remove from such companies as IBM.

After 2006, Huawei began to generate more and more orders from overseas. As monetary values rose dramatically, Huawei's financial management capabilities were unable to control end-to-end processes for large projects, as IBM and other large companies could do. This led to problems with the profitability of many large projects and increased loss of control over cash flow.

Huawei was signing quantities of contracts with customers in Europe and Africa, for example. Once the equipment was handed over, however, and final reckoning was made of payment, the company was discovering that some contracts had made no money at all or were borderline. This put Huawei up against a bottleneck in further growth. If the company could not achieve a globalized financial oversight system, it would not be able to safeguard net profitability and cash flow. The more ferociously it expanded overseas, the more it put itself at the risk of operating losses.

In mid-2008, Huawei began to institute new procedures to improve the operating efficiency of its global sales effort. It worked together with a company called Accenture to implement customer relations management solutions with its more than 300 key customers around the globe. This was to help customers reduce their own operating costs and improve their own competitiveness via providing outstanding products and services that met their needs. Not only did this "turn opportunities into contracts, and turn contracts into cash," but it also vastly strengthened profitability. The speed with which cash turned over increased dramatically.

In 2008, Huawei realized annual contractual sales income of USD 23.3 billion, operating income of USD 17.5 billion, and net profit of USD 1.15 billion. Its operating profit margin went from 10 percent in 2007 to 13 percent in 2008. In terms of the growth rate of sales to the growth rate of costs, it achieved the high ratio of 3.4 percent. Meanwhile, its expense ratio fell from 28.5 percent in 2007 to 26.7 percent in 2008.

Huawei also began to implement what it called an "iron triangle" in terms of how it structured its organization in overseas markets. It split up the previous nine large regions into 22 regional departments that were closer to frontline markets. This enabled it to set up forces that were "within sight of the cannonballs." It also set up decision-making structures that were similar to those employed by the U.S. Joint Chiefs of Staff. Combat forces or "units" under each of the 22 regions included the "iron triangle" of customer managers, product managers, and order fulfillment managers. After discovering an opportunity, these teams could make on-site decisions within their authorized scope of activities. When they encountered major projects, they could rapidly call for reinforcements from regional headquarters. This greatly reduced the costs of internal communications. It improved the timeliness of decision making. It allowed for quick reactions in terms of "locking corporate jaws onto opportunities." All this improved Huawei's fighting ability swiftly and dramatically.

Advancing into the World 500

This series of reforms within Huawei enabled it have a massive combat force that stayed close to frontline customers. It also allowed for a substantial reduction in management costs, which in turn enabled the company to provide customers with an outstanding price-to-functionality ratio for products and services. When the financial crisis erupted in 2008, the primary beneficiary of that crisis was therefore the already well-disciplined corps of Huawei people.

Huawei was well positioned on the Asian battlefield when China issued three licenses for 3G business in early 2009. The company already had an absolute advantage in the world's most complete product line, which now incorporated the three major formats of 3G. It had all its wishes fulfilled within the China market when it broke through the figure of USD 10 billion in sales.

In India, Huawei had set up its first overseas R&D Institute in Bangalore in 1999. After 10 years of diligently cultivating these fields, by 2009 this huge market, second only to China, also produced a bountiful harvest. In 2008, Huawei replaced North American firms in taking over GSM orders worth USD 1.7 billion. Then, on May 10, 2009, India's largest telecom operator, BSNL, announced that it was dividing the world's largest order of 3G equipment, valued at USD 6 billion, between two providers, Ericsson and Huawei. Meanwhile, in Japan, Huawei's 3G and end-user products broke one by one into the markets of Japan's largest operators. First came Japan's number one operator, NTT DoCoMo, followed by its third-largest operator, Softbank. Then, at the end of 2009, Huawei also successfully broke into the markets of Japan's second-largest operator, KDDI. Huawei's 2009 sales in Japan exceeded USD 400 million.

On the European battlefield, following on the "red" commotion Huawei had stirred up in both eastern and western Europe, events in northern Europe now also moved in the company's favor. In June 2009, northern Europe's telecom operator Telenor announced that it was choosing Huawei to deploy the world's first 4G commercial network throughout Norway. In November 2009, Finland's operator TeliaSonera announced that Huawei would be setting up a 3G network within the country.

At the end of 2009, the Swedish operator Net4mobility announced that both Huawei and Ericsson had won bids to implement a 4G network project in that country.

In the short space of six months, Huawei successfully broke into three northern European markets in succession and thereby pierced the final "Maginot Line." This meant not only that Huawei had overtaken Nokia and Siemens but also that it was now beginning to stand on an equal basis with Ericsson in the realm of global wireless technology.

The most astonishing news, however, came out of North America. On March 30, 2009, the U.S. operator Cox announced that Huawei would be the solutions provider for its end-to-end CDMA mobile network. Before this, Huawei and a mobile operator named Leap Wireless signed contracts worth USD 100 million to deploy 3G networks in Washington State, Idaho, and Nevada. Canada's operator Telus and Bell Canada now jointly announced that Huawei would be constructing a portion of Canada's 3G and 4G networks under contract. Although none of these three North American contracts was large, they signified that Huawei was finally overcoming the most entrenched barriers to entry in the world.

In April 2005, *Time* magazine chose Ren as one of the "100 most influential people on the globe" in that year. He joined the ranks of such IT giants as Bill Gates, Steve Jobs, Larry Paige, and Sergey Brin. In December 2008, *BusinessWeek* chose Huawei to be among the 10 most influential companies on the globe, putting it in the ranks of such companies as Apple, Google, Unilever International, Walmart, and Toyota.

In February 2009, the World Intellectual Property Organization (WIPO) announced that Huawei had submitted applications for 1,737 patents in 2008, which made it the global champion in terms of numbers of inventions, beating out Matsushita, Phillips, Toyota, and others in the process.

In July 2010, *Fortune* magazine published its Fortune 500 for the year. Not only did Huawei join the list as 397, with its USD 21.8 billion in operating revenues, but also it joined without having to be listed on a stock market. It was the one and only company on the list that was not publicly owned.

The world's major media were quick to follow up on this. An editorial in the *Washington Post* declared, "Huawei, located in Shenzhen, has become one of the world's strongest telecom equipment manufacturers. With its patents and innovations, it has become a symbol for China's new-style enterprises, but it also has become a world leader in innovation." Meanwhile the consulting firm Ovum, which focuses exclusively on telecom R&D, made the point in a special report that "Huawei has transitioned from being a Chinese vendor to being a global giant."

In terms of the structure of the global market, Huawei now possessed an absolute advantage in Africa, Latin America, and the Asia Pacific, where it held about 40 percent of market share. It also held more than 30 percent of market share in Europe, after surpassing Alcatel and Lucent and Nokia and Siemens in the 3G arena and coming neck and neck with Ericsson in the 4G arena. In 2008, Huawei went from providing services to 36 of the 50 largest operators in the world to providing services to 46 of the 50 largest operators. The only ones the company had not yet conquered were AT&T, Verizon, Sprint, and T-Mobile. However, these four American operators constituted more than 35 percent of the global procurement volume for telecom equipment.

Given the importance of this market, Huawei now began to do everything it could to penetrate it.

First, to set up beneficial relations with American customers and develop products they needed, Huawei set up R&D centers in Dallas, Texas, and Silicon Valley. It recruited large numbers of professionals from around the world, including senior management as well as R&D staff. The former chief technical officer of British Telecom was among them, Matthew Bross, as well as the former chief technical officer of Nortel, John Roese. To counter the negative impression that the American government had of Huawei, and to moderate concerns about security, Huawei set up an office on K Street in Washington, D.C. It hired a number of senior former U.S. government officials in order to create a powerful public relations team to carry out direct and open promotion and lobbying efforts. To express good faith, the company voluntarily provided original source code for its equipment to a professional evaluating team appointed by the U.S. National Security Commission. Equipment could be sold on to Huawei customers only after this body had inspected and approved the equipment. The sale was done via a third party, which purchased the equipment from Huawei and resold it to the American customer. Despite such massive efforts, however, Huawei still found itself far from being able to realize its American dream.

After Huawei teamed up with Bain Capital in a failed attempt to purchase 3Com, in July 2010, the company participated in a tender for Motorola wireless equipment assets through a company it had purchased for USD 2 million, called 3Leaf. Nevertheless, in the end, it still had to declare defeat.

In November 2010, Huawei finally won a tender from America's third-largest operator, Sprint, with a bid for USD 5 billion. Huawei came first in terms of equipment evaluation, technology, and business terms. Nevertheless, at the critical moment it ran up against U.S. national security considerations and was again taken out of the action before the deal was consummated.

On May 5, 2014, Ren was interviewed by journalists in London. A reporter from the *Daily News* named Chris Williams asked him, "Do you think that the challenges you are facing in America stem from a lack of trust or from trade protectionism?"

Ren responded as follows:

America is a very great country. It is an advanced country, with flexible mechanisms and clearly defined property rights. These provide powerful support when it comes to respecting and protecting individual rights. These tremendous mechanisms, situated in an open kind of culture, are what have enabled America to keep its leading position over a long period of time. Because of this, we have never wavered in our desire to learn from the United States. During the 9.11 incident, when planes bombed into the World Trade Center and when people were frantically trying to escape, those rushing downstairs made way for firemen rushing upstairs. People were courteous in making room for those who were disabled. The situation was extremely orderly. This kind of spirit is not something that forms overnight. This caliber of culture comes out of a process that gestates over several hundred years. All of these things contribute to America's greatness. Whether or not the United States buys Huawei's equipment is inconsequential next to these things. Compared to these bigger things, how Huawei is treated is a small thing.

CHAPTER FIFTEEN

TAKE IT TO THE LIMIT

We are like a winged steed flying across the wide open steppe.
Nothing can stop us, except for our own indolence and our
own corruption.

—REN ZHENGFEI

A meeting of Huawei's board of directors was held in Sicily on January 17, 2011. At this meeting, Ren Zhengfei set forth a more farsighted goal than he had ever presented before.

"Within 10 years, Huawei not only plans to be a leader in technology, but it plans to be running neck and neck with such western tech giants as IBM, Cisco, and Hewlett-Packard, with annual revenues of USD 100 billion."

This declaration was undeniably a challenge to the top tier of global corporations. One should realize that the figure of USD 100 billion is roughly equivalent to New Zealand's total GDP. Very few IT companies in the world come close to this figure, although they do include IBM and Apple. Ren's ambition and his determination to "take it to the limit" were astonishing but they also raised doubts. People might believe that Huawei could produce yet another astonishing coup, but they also recognized that going from USD 30 billion to USD 100 billion was no small undertaking. Not only did the company face brutal global competition in terms of its external environment, but also it needed to restructure itself internally. It needed to transform its operating methods and build up its corporate culture. The question therefore became, What arrangements has Ren made to deal with these things, and what issues still remain? Growth was slowing down, the organization was aging, and China's advantages in low-cost labor and being a tech follower had peaked. Could Huawei really escape the fate of most Chinese companies and keep from declining after such a swift rise? All this was definitely going to be an ultimate test for Ren.

Moving Forces into New Battlefields

Over the past 27 years, Huawei had consistently kept its feet planted in telecom markets. It had achieved massive success by doing this, and 95 percent of its operating revenues came from this sector. As the peaks of installing 2G, 3G, and 4G went by in succession, however, and as ongoing changes in Internet technologies impacted business, telecom markets were becoming saturated. Meanwhile, the past several years had seen absolute bloodbaths, accompanied by several intense spates of mergers and acquisitions in the industry. Twelve main telecom equipment manufacturers had been reduced to just a few. The survivors included Ericsson, Alcatel and Lucent, Nokia and Siemens, Zhongxing, and Huawei. Growth rates and profit margins were both now highly limited. If Huawei was to achieve its objective of USD 100 billion, it would have to consolidate its leading position in telecom markets and it would have to open up new horizons and become the entity "pushing the wave" in core technologies. To progress from being a telecom equipment manufacturer, it would have to achieve a strategic transformation to being a full-capacity IT provider. This was imperative and the situation required action.

In early 2011, therefore, Huawei broke out of its old mode of operating. It restructured a business that had been organized according to customer groups and instead organized itself according to three large business departments, namely, those dealing with operators, enterprise networks, and end-users. The company began to move its forces into broadly based areas that lay outside pure telecom markets.

The business in enterprise networks not only was huge, but had respectable profits as well. Nevertheless, it had long been occupied and indeed monopolized by the top global giants, who had massive strength and also were in control of core technologies. Other than storage and server products, whose markets had already been carved up by IBM, Hewlett-Packard, and Dell, the main other area was data transmission. Huawei was relatively strong in this area, but Cisco already held more than 80 percent of market share worldwide. It was going to be extremely difficult for Huawei to occupy this already mature market.

In the consumer market for end-user cell phones, Apple was the leader in high-end smart phones and was guiding the trends as this business developed. South Korea's Samsung had a certain degree of brand influence. Huawei's new forces were coming to the fore with admirable results. Nevertheless, if the company intended to compete with global leaders on a long-term basis in this area, its R&D would have to surpass the core technologies of Apple and Samsung. For Huawei, this would require an extremely tough assault on heavily fortified positions.

Huawei's rotating CEO, Xu Zhijun, commented on the situation. When interviewed by *Fortune* magazine, he said, "Huawei's business was very focused on operators before. Now, it is attempting to broaden that out and is trying to be geared not only toward operators but also enterprises and consumers. This represents a monumental challenge." Clearly, if Huawei intended to achieve the goal of USD 100 billion, the company would have to carve a bloody path for itself through the "Red Sea" of markets beyond just telecom, and it would have to transform itself from being a tech follower to being a tech leader.

Setting the Pace in Cloud Computing

Cloud computing is an emerging technology that, in simple terms, obviates the need for such things as traditional data centers, servers, and computers. All that companies or individual consumers need is a small end-device, a piece of equipment that can connect to "the cloud." What that means is that the end-user is connected to a cloud service provider. This allows the end-user to obtain anything it might need in the way of information services, including software, computing, technology, and storage. As with the Internet, the birth of cloud computing will change the entire information industry. Because of this, it has been dubbed the "third revolution in IT." IBM, Cisco, Google, and Amazon are therefore putting massive amounts of investment into R&D in this area, in order to be the ones who guide the development of this new wave in the future.

Ren naturally did not and does not intend to miss this chance to go from being a follower to a leader, since it represents a historic opportunity. At the end of 2010, Huawei formally announced its strategic plan for cloud computing in Beijing to its global customers. It also unveiled its end-to-end product solutions. In the press conference, Ren made a high-profile announcement about the company's future development plans.

"In terms of cloud platforms, we intend to overtake Cisco in a fairly short period of time. In terms of cloud services, we intend to go after Google. We aim to provide all people in the world with the ability to enjoy information applications and services just as though they were using electricity."

"Huawei has already arrived at the cutting edge of the telecom industry, so it is harder to determine where our next step should go. In the past, we relied on western companies to lead the way. Now we ourselves must participate in that process. As western companies do, we must work hard to make a contribution to the world."

To have a head start at the starting line of cloud computing, between early 2011 and late 2012, Huawei recruited 30,000 new employees. For the first time in its history, the company employed more than 130,000 people. Among these, almost 60,000 were distributed among the domestic research centers in Shenzhen, Shanghai, Nanjing, Hangzhou, and Xi'an, where they carried out R&D. At the same time, in order to broaden its international perspective and expand its R&D strength in North America and Europe, starting in 2011, Huawei set up joint-venture innovation centers with Canada's mobile operator Telus and with Canadian Bell. Huawei will invest a lot of money to acquire 49 percent stake in a joint venture between Huawei and Symantec. After seeing the movie *2012*, Ren made the comment that Huawei had better start thinking about how to guide the "information flood" in the information explosion. It therefore might want to think of building a kind of Noah's Ark. To that end, Ren began strengthening Huawei's basic research in cloud computing. In May 2012, the company established a Noah's Ark Laboratory in Hong Kong, at the Hong Kong Polytechnic University. It hired a number of Chinese and foreign cloud computing experts, who began to conduct in-depth R&D. On July 2, 2012, Ren described his strategic thinking on this to people working in the laboratory as well as experts who had come together for a forum.

"The purpose of this Noah's Ark Laboratory is to deal with the impact of the information flood. It is to find solutions to the various challenges posed by collecting, organizing, mining, and analyzing information. It is to structure Huawei's technological advantages so that it becomes the leader in the information and communication technology (ICT) arena. In order for our own ship to overtake and surpass its 'naval escort,' we must have new ways of thinking and we must generate new theories. In sum, this Noah's Ark must serve Huawei as the ship that navigates for us in the midst of this information flood."

Ren also expressed his thinking about alliances with others in this process. This involved the intersecting considerations of politics and business. It related to the need to maintain friendships with "all under heaven" in the new and highly complex global situation, which meant figuring out how to create win-win connections in the overall "ecosystem." Ren was characteristically succinct. His solution: "openness, cooperation, and win-win." This also would allow the company to transform itself from being a follower to being a leader in the cloud computing age, as it integrated the sum of humanity's knowledge into its systems.

"If a culture is not open, it is impossible for it to draw in the best features of others. A culture that is not open then gradually becomes marginalized and reaches a dead end. Sooner or later, an organization that is not open turns into a stagnant pond, since not being open is equivalent to being dead."

At this juncture, Huawei finally began to crack open gates that had long been closed to it in the outside world. It began to communicate with an open attitude and to connect with others. Huawei's senior collective leadership began a series of "image" initiatives, both on the company's open blogs and in various forms of media. The message was that "Huawei is more than simply one of the World 500." Xu Zhijun, Guo Ping, Hu Houkun, and other rotating CEOs and senior officials, including Chief Financial Officer Meng Wanzhou, agreed to interviews with the Chinese edition of *Fortune* and other well-known publications. They gave detailed responses to issues in which the outside world was particularly interested, such as Huawei's cash-flow situation, its succession strategy, and the possibility that it would list on the stock exchange.

On the afternoon of May 8, 2013, Ren received four media representatives from New Zealand and gave the first formal interview that he had ever granted to anyone. He answered questions that ranged from Huawei's R&D to its international strategy, to becoming a leader in global technology in the cloud computing age.

All the above is a reflection of Ren's ambition and determination to lead the way in cloud computing. Right now, we cannot imagine how Huawei will actually navigate its Noah's Ark through what are bound to be very turbulent waters. Nevertheless, one thing is certain: Huawei has taken the first heartening step in the direction of realizing its dream.

"Compromise and Shades of Gray"

Between 1998 and 2008, Huawei completed four major management reforms in the areas of R&D, supply chains, finance, and customer relations. These were achieved in part by incorporating IPD and ISC into the corporation in 1998 and Integrated Financial Systems (IFS) and Customer Relation Management (CRM) in 2008. Once these were assimilated, Huawei cast off the habitual assumption that a Chinese company is just a "sales organization." Instead, Huawei set up operating procedures that were customer-centric throughout the enterprise.

The analogy could be made to a dragon, with frontline markets being the dragon's head. Once business objectives and customer needs are identified, this dragon head begins to move and its movement then causes sympathetic oscillation all along its body. Each joint, each organ, begins to sway in response. Soon the entire dragon is dancing along.

In setting up 22 regional operating departments around the world, Huawei also carried out what it called a "beehive" restructuring of the organization. Responsibility for any given "item" in the beehive was propelled by the business value of that piece of business. Any place with flowers determined the orientation and the "battlefields" of business. The beehive could swiftly come together of its own volition. Once it had collected its nectar, it could again disperse, to wait for the next opportunity to assemble.

These things represented the professionalism that Ren encouraged when he said, "Costs are lowest when you are doing the same thing at the same time under the same conditions." Once Huawei mastered highly efficient business procedures and organizational structure, it could wield its innate advantage in low-cost labor and take the initiative in being competitive in foreign markets.

After years of exploratory trials, importing such western management systems as IPD, ISC, IFS, and CRM did indeed play a major role in elevating the company's competitiveness in international markets. Nevertheless, there was always going to be a certain cultural disparity between east and west. In the process of becoming international, the trick for Huawei was to master western management procedures while at the same time taking advantage of eastern techniques. That is, it had to learn to discipline itself in western boxing while also being proficient in Chinese *gong-fu*, and in such a way that it developed a more nimble and professional team than could exist in western corporations. In this regard, Ren espoused a management philosophy that he called "compromise and shades of gray."

Western professionalism has summed up lessons from more than 100 years of a changing market. That is why it is so effective. However, you don't necessarily have to put on a coat and tie to look good. When we emulate that, we should not necessarily copy it over wholesale, be rigid about it. After all, why shouldn't it be all right to wear a Chinese-style suit? After 20 years, we have our own successes and we want to sum these up as well. We want to make use of them, figure out why they were successful and then keep on doing them. In addition, we want to adopt those western methods that make our own successes more standardized and routine, more of a baseline for how we operate. Instead of just copying western models, we want to have an effective corporation that manages with a living soul.

The important qualities of a leader include his orientation and his rhythm or cadence. Both of these should have an appropriate amount of uncertainty. A clear direction is necessary in the midst of murky conditions, to stand out against a gray background. However, orientation is something that changes, depending on time and space. It often becomes less clear, but that does not mean that it has to be black if it is not white. It is not a question of just this or that. Leaders should have a grasp of the right amount of gray. This allows all considerations in any given period to achieve harmony as they influence the forward momentum. This harmony is what could be called compromise. The result of this kind of harmony is what could be called a certain shade of gray.

Compromise does not imply that you must abandon your principles. It also does not mean blindly giving way. Intelligent compromise is an appropriate exchange. Once you have reached your primary objective, you can give a little, to the right extent, on less important objectives. Instead of regarding this kind of compromise as a loss of principles, it should be seen as using retreat in order to advance. Through the right exchange, one can accomplish the sure realization of one's objectives.

An unintelligent compromise, on the contrary, represents the loss of appropriate balance. Holding too tightly to lesser objectives can mean losing your primary objective. Or the price you pay for compromise can be too high and can lead to unnecessary loss. Intelligent compromise is an art in how to give way. Compromise is a kind of virtue, and is one of the qualities that a manager must possess.

No river on earth is straight. All have curves. They represent the compromise between rivers and earth. Such compromises are what allow the waters of a river to reach the ocean.

Ren's philosophy of "compromise and shades of gray" is extremely rational and pragmatic. It is a kind of soft power, and it will be necessary for Huawei's managers and employees to learn to use this soft power if Huawei is to become a great corporation. Making use of this philosophy is a kind of threshold that Huawei must cross over if it is to use its soft power to win victory in the contest with the giants of the world.

Struggle is Fundamental

"Huawei will become a core player in whatever industry it decides to go into." This statement was made by a member of the press who was astonished at Huawei's transformation from a small workshop operation to a massive telecom giant. It also accurately describes the way in which Huawei people have created one astonishing success after another over the course of the past 20 years, based on unrelenting effort. Nevertheless, after so many years of sustained battle, even the super strong physical strength of Huawei employees has begun to "build up an overdraft," borrow against the future. Huawei people are in fact now coming up against both physiological and psychological crises.

To expand markets, Huawei consistently held to its principle of "outstanding quality, excellent service, and low cost." Customer needs have guided growth strategies. In the past, its own technologies, products, and brand influence were far behind those of its opponents. In the midst of intense international competition, all Huawei people could do to succeed was work "through the coffee breaks," work harder than others. As a result, they sacrificed their own health more than others did in order to achieve the goal of overtaking others.

In 1996, the head of Huawei's publicity department, Zhu Jianping, recommended putting out a newspaper to enrich the extracurricular lives of Huawei employees. Ren firmly supported this, but the name that he himself gave to this paper was *Outside of 24 Hours*, which basically meant that any individual activities and personal pleasures had to be conducted outside of the 24 hours a day that people were to devote to Huawei. As Ren himself often said, "Coming into Huawei is like entering a tomb." The work pressure and psychological pressure on people were extreme. Lights were generally on in Huawei buildings way beyond 10 p.m., and not just at headquarters but at the company's 110 branches around the world. The pace of work was ultimately too much. Around 2008, the health of some cadres and workers got affected. To this problem, Ren frankly agreed that while Huawei's compensation was higher than average, the company had failed to provide assistance with the mental well-being of employees. In August 2008, he made a speech encouraging people to take an active, open, and straightforward approach to life, and he used his own personal experience as a way to connect psychologically with employees. He called this speech "Going through life's difficulties with a positive attitude."

"...there is nothing more beautiful than life during our time on earth. We must value life and protect it. When people are too tired, they must rest."

In the second half of 2008, the company therefore created the new position of "chief officer of employee health and well-being." One of the company's founders, Ji Ping, was assigned to this position. He soon set up a number of systems that elevated health and well-being to the same status as employee training and employee benefits. He also set up a "mental health advisory team" and a "health guidance center." These delivered information on personal health to employees and how to strike a proper balance between work and rest. They promoted safety measures in this regard. Between 2008 and 2012, Huawei spent a total of RMB 1.9 billion on safeguarding employee well-being.

Nevertheless, as the company advanced its forces into new markets and as orders flew in like proverbial snowflakes, it was impossible for employees not to work overtime. What's more due to Huawei's rise in the international arena, its competitors also now began investing more in R&D and trying to upgrade the level of customer satisfaction. They took all possible measures to counter Huawei's swift and fierce attack. For example, after Huawei became the world's second-largest telecom equipment manufacturer in 2009, the average work week at Alcatel and Lucent, Nokia and Siemens, and Ericsson went from an average of 40 hours to an average of 70 hours.

Maintaining a lead in technology was obviously going to be critical for survival in an industry engaged in brutal competition. This meant it was impossible for anyone to avoid a hard-and-fast work pace, but particularly Chinese companies, which had to "catch up with and then overtake" more advanced countries in Europe and America. They had to work harder and faster if they were to take a seat on the global stage of high-tech. Not only did Chinese companies have to work night and day, but also they had to maintain both physical and mental discipline.

Ren addressed this subject in an article published in the company's internal magazine, called "Heaven rewards diligence."

"Some people may think that what Huawei has achieved is already impressive. Some may even think that our 'cot culture' is no longer appropriate. We ought to relax a little. I believe it is dangerous to think in this way. Crises lurk right behind prosperity. These crises are not necessarily features of prosperity itself. Instead, they lie in the mentality of people who are immersed in prosperity."

"Bitter struggle will inevitably achieve prosperity, but if that is not followed up with ongoing struggle, prosperity will inevitably be squandered." He then quoted a line about how the rise and fall of events have rolled on from time immemorial, like the Yangtze River. He noted that history is like a mirror and provides plenty of instructive examples. "We must maintain our struggle over a long period of time or we will die out, and the most important part of our struggle is mental. At every moment, we must maintain a sense of crisis. We must preserve a cool-headed approach to our success. We must not become arrogant."

"Intense struggle is the very soul of Huawei's culture. It is our main theme, something we cannot relinquish for a second. No matter how large and powerful Huawei becomes, we cannot let go of this fundamental attribute."

To inspire the fighting spirit of employees and keep up Huawei's enormous cohesion, strength, and ability to execute its mission, Ren then initiated a major discussion about Huawei's core values that went on for over two years. It resulted in the determination that the corporate culture and its core values involved "taking customers as central and also regarding struggle as essential in the new period."

To that end, for the first time the company now revealed on its internal website that Huawei had paid more than RMB 100 billion in taxes to the state between the time the company was founded in 1987 and 2010. Huawei also began to implement incentives that were in line with the idea of "being customer-centric and regarding struggle as fundamental." These leaned in the direction of providing more shares and more salary and bonus benefits to those "who fought for them." They also gave preferential consideration to "those who struggled" when it came time for promotions and the cultivation of managers who would be in line for succession to senior positions. Huawei now put a larger number of people on the Huawei board who had relied on their own "struggle" to achieve outstanding results.

Meanwhile Ren also undertook a key measure to put an end to the phenomenon of corruption and the idea that employees should look after their own interests. That is, he implemented what was called an "EMT self-discipline oath." Starting in 2008, the company began to hold a meeting once a year in which the CEO Ren, the chair of the board, Sun Yafang, and core members of the EMT team lined up, raised their right hands, and swore a declaration to several thousand managers and employee representatives. This oath included such statements as "We love Huawei as much as we love our own lives." "We will never abuse the authority that Huawei has bestowed upon us to influence or interfere with company business in ways that seek self-interest." Senior management then declared its willingness to be monitored by the entire body of employees, in terms of accepting supervisory oversight.

This procedure stemmed from a recognition that Huawei's resources derive only from the minds of those who fight on behalf of the company. They are its timber, its coal, its oil fields. While the force of financial benefit propels the performance of these people to a degree, the wellsprings of superlative performance come from the example set by senior management. Senior managers provide the role model for selfless, fearless, clean, and upright behavior and for an indomitable will to succeed.

Moving in the Direction of Listing on the Market

In 1988, Ren made a one-time purchase of 200,000 shares of the company Vanke. In 1995, when Vanke's shares were trading at an all-time high, he sold those shares and made a considerable profit. With the money, as well as retained earnings that the company had amassed over years, he purchased a piece of undeveloped land on the outskirts of Shenzhen in a place called Bantian Village, in the Long'gang district. He was able to buy this 1.3 square kilometers of raw land for a fairly low price. As Huawei rose in prominence, other companies began putting corporate offices in the same area, including Xintianxia, Fushakang, Biyadi, Rilihuanqiu,

Oumulong, and Saigesanxing. By now, Bantian has become a modern industrial park and land values have soared.

This purchase and sale of Vanke shares, a first "testing of the water," provided Ren with more than simply capital. Through this contact with shareholding systems, he began to understand the unique attraction of shareholding and how the system could work in Huawei's favor. In Huawei's early period, compensating employees with shares was highly successful in motivating people and propelling innovation. It enabled the company to draw in outstanding talent and thereby to strengthen its business.

The shareholding system of Huawei helped create Huawei, but the company has not in fact listed on the market. Not only is this unique within China, but also it is rare around the world. On the list of World 500 companies in 2009, a total of 499 were listed on markets. Only Huawei was not. Why has Huawei not gone public, and how does it derive its operating funds without listing?

As described above, when the IT bubble burst in 2000, almost taking Huawei down in the process, Ren made preparations to list within China as a way to prepare for a rainy day. He later gave up this idea for a variety of reasons, including national regulations which at the time stipulated that no more than 200 shareholders could own shares if a company was going to be listed.

In 2001, Ren began to consider listing outside of China. He drew up a plan for bringing in five or six strategic investors by increasing shares by 30 percent. A newly constituted board would then list overseas. After several rounds of discussions with large international companies on the subject, however, including IBM, Intel, and Marconi, he again tabled the idea.

Since these two potential listings came to nothing, Ren now considered spinning off product lines as a way to generate cash flow and accumulate funds. He adopted this innovative method of financing in 2001 when Huawei sold its electric power business for USD 750 million and again in 2007 when it sold 51 percent of shares in H3C to 3Com for USD 908 million. These were just two examples of how Huawei solved its need for funds. As a result, the company's operations and decision-making mechanisms also became more nimble, more able to respond quickly when opportunities arose. This made it unlike listed companies, which often lost out due to a prolonged decision-making process. In addition, Huawei was not required to divulge business results to the public. It therefore could focus its core efforts on expanding overseas. In a sense, not listing on the market is part of what has "created" Huawei.

By now, however, as the company faces increasing international competition and as it seeks to undertake a strategic transformation, listing on the market appears to be a necessary course of action.

First, such world-class companies as Microsoft, Cisco, IBM, and Hewlett-Packard have all become what they are as the result of constant mergers and acquisitions. If Huawei intends to become equally powerful, it too must consider the issue of acquisitions. Second, as traditional telecom markets move in the direction of being saturated, Huawei will unavoidably be in competition with powerful giants backed by deep resources as it internationalizes. All these are aiming for markets in enterprise networks, end-users, and cloud computing. Not only will Huawei need to invest massive amounts in core technology to go from being a tech follower to being a tech leader, but also it will need to ensure adequate cash flow to reach its key objective of USD 100 billion in revenues. One option would be to spin off portions of the company and list them overseas. Huawei could also bring in strategic shareholders. Per international practices, it could expand its reputation and brand influence in enterprise networks, end-user markets, and cloud computing through strategic mergers and acquisitions. Through such measures, it would also obtain long-term, stable funding support.

Preparing the Public for a Rotating CEO

On April 3, 1999, Ren took questions from newly trained employees in press conference style at the famous holiday resort in Shenzhen called Green World. These employees had just finished their systematic all-round training and were now setting off to be frontline sales staff. One of them asked respectfully, "Ren-*zong*,[1] what will Huawei do after you retire with honors?"

---

[1] The suffix –zong is an expression of respect. It is often used for someone who heads a company or an organization.

Ren responded, "Actually, I personally have not contributed that much. The ones who deserve the honors are our mid-level and senior management as well as the entire body of Huawei employees. Not only do I not understand technology or IT, but I can't even read financial statements properly. My sole advantage is that I am willing to change when I make mistakes, since I am not overly concerned about face. That attribute is not so hard to find in a person, so I feel that finding a successor won't be much of a problem. The next person just needs to be fairly democratic, and able to sign things. What we must avoid at all costs is mythologizing any individual. If that happens, we distort the systems by which Huawei creates value, and that will take the company down."

Ren was 55 years old when he made this comment. Since he was in the prime of life, talking about succession seemed premature. Time shows mercy to no man, however. Over the next 16 turbulent years, Ren was to achieve his dream of making Huawei one of three main players in the telecom business, carving up international markets with European and American giants, as well as his dream of taking Huawei into the ranks of the World 500. Along the way, no successor appeared. This person "who could sign things" and who "was fairly democratic" did not come forth. By now, this has become the focus of considerable attention outside Huawei.

As described above, once Ren had founded Huawei, his personal charisma derived in part from the way he consistently used his own behavior as a model for others. Such behavior included a selfless approach to contributing to the company, and it had the effect of creating an unassailable degree of authority. This was invaluable in taking the company through its difficult start-up period, when it was surrounded by a fairly vicious environment. Ren used a superlative sense of entrepreneurship and also pure "guts" to take Huawei from the status of small workshop to a large consolidated corporation.

By the second half of 2003, the implementation of IPD, ISC, and other western management systems were beginning to take root in the company. Two considerations now came to the fore, namely, setting up group decision-making mechanisms and cultivating a successor. Based on these, Ren worked with IBM consultants to set up an EMT operating management team. Specifically, they set up a system of rotating CEOs which has evolved over time and is quite unusual. Eight members of a collective decision-making team serve as CEO for six months at a time on a rotating basis. Each person rotates through two terms, during which he or she is in charge.

This succession model has been able to take advantage of collective intelligence. It has reduced the extent to which one individual can act arbitrarily without consulting others, and therefore the extent to which the company is stymied by mistakes and prejudices. At the same time, it has avoided the enormous risk to a company's "fate" when too much power is concentrated in one person's hands and he follows the same set of procedures that were always followed before.[2] Both strategy and the hiring and firing of senior managers are decided together as CEOs rotate. After each CEO finishes the six-month term in office, she or he does not leave the core management tier, so continues to have considerable power and prestige with respect to the execution of strategy, business decisions, and the appointment of senior managers and employees. This prevents each new administration from having its own slate of officials. It allows for stability and continuity of job positions under rotating CEOs. It also puts a stop to the problem of having any one person hire relatives and friends when such hires are not appropriate.

This rotating system of CEOs is something that Ren has put years of thought into developing. It is a "group succession model" that has been the result of considerable exploration and trial efforts. As an innovative succession model, it is absolutely unique in corporations around the world. Whether it will enable Huawei to be one of those enduring old-name firms is of great interest to people. Naturally, the only thing that does not change in the world is change itself. In recognizing this truth, Ren is candid when he says, "If it isn't successful, at least we will have explored the path ahead for those who follow us."

With regard to the essential nature of "succession," however, he also has emphasized many times in internal meetings that "Huawei's succession is not a person-to-person changing of the guard in a feudal fashion. CEOs of western companies circulate constantly, like those horses in a

---

[2]The text uses a phrase here that relates to a Han-dynasty example, that is, an example from 2,000 years ago.

paper carousel lantern at New Year's. Yet I haven't noticed that western companies suffer for it. American presidents turn over constantly as well, yet America still stands. What we want to achieve in Huawei is therefore the ongoing succession of a pattern, a culture, a system, a set of procedures. We do not want to think of succession in terms of just one person. In the future, no matter who is running the company, there should be no change in our core values."

All this notwithstanding, as Ren's retirement becomes increasingly imminent, one has to wonder whether everything will proceed according to his vision of the future.

He himself has said, "History moves in cycles and calamities always come 'round again. Meanwhile, calamities have never yet changed human nature and human greed. Will we be able to save ourselves if the world changes around us? Keep ourselves in balance if the financial crisis explodes again, or we have a radical devaluation of the currency, or our society becomes highly unstable? Will we have people who, like Koreans, are willing to sell their own jewelry to buy fuel for the common cause? I wonder. History itself has no endpoint. How, then, can we imagine that our good fortune will last forever?"

Right now, the battle is already joined as Huawei begins to compete against such global giants as Cisco, Hewlett-Packard, Apple, Google, and IBM in enterprise networks, end-user markets, and cloud computing. As Huawei ascends one peak after another, it is finding that the air is thinner, the oxygen more scarce. It is also finding that the chasms on either side of it are more precipitous and daunting. Nevertheless, we firmly believe that no matter how tortuous the path that lies ahead, the 190,000 "successors" that Ren has cultivated will be able to shoulder the burdens. They will take on the responsibility of making Huawei a leader in the global telecom industry. They will continue to make new contributions to advancing science as well as advancing all humankind.

INDEX